Miller's Guide to Home Plumbing

Miller's Guide to Home Plumbing

GLENN E. BAKER
Professor Emeritus
Texas A&M University
College Station, TX

REX MILLER
Professor Emeritus
State University College at Buffalo
Buffalo, New York

MARK R. MILLER
Associate Professor
Texas A&M University–Kingsville
Kingsville, TX

McGraw-Hill

New York   Chicago   San Francisco   Lisbon   London
Madrid   Mexico City   Milan   New Delhi   San Juan
Seoul   Singapore   Sydney   Toronto

Library of Congress Cataloging-in-Publication Data

Miller, Mark R.
    Miller's guide to plumbing the house / Mark R. Miller, Rex Miller, Glenn E. Baker.
        p.   cm.
    Includes index.
    ISBN 0-07-144552-8
        1. Plumbing—Handbooks, manuals, etc.   2. Dwellings—Maintenance and
repair—Handbooks, manuals, etc.   3. House construction—Handbooks, manuals, etc.
I. Miller, Rex,   II. Baker, Glenn E.   III. Title.

    TH6125.M54   2004
    696'.1—dc22
                                                                    2004058785

1 2 3 4 5 6 7 8 9 0   QPD/QPD   0 1 0 9 8 7 6 5 4

ISBN 0-07-144552-8

*The sponsoring editor for this book was Larry Hager, the editing supervisor was
Caroline Levine, and the production supervisor was Sherri Souffrance. It was set in ITC
Century Light by Wayne A. Palmer of McGraw-Hill Professional's Hightstown, N.J.,
composition unit.*

Contents

7 Major Plumbing Projects

8 Rural Water and Waste Systems

Preface

Miller's Guide to Home Plumbing was written by the authors of McGraw-Hill's very popular *Carpentry and Construction,* in publication since 1981. This work on plumbing is written for the homeowner and features several unique sections. It covers almost anything a homeowner may encounter, from simple plumbing repairs to new installation. It starts with a broad overview of what plumbing is and some of the conditions and laws that affect it. From this it leads to a comprehensive review of plumbing codes, followed by factors in planning a plumbing job.

Next, sections on the various general and specific tools needed for either simple or advanced plumbing jobs are presented. Ideas on what to purchase or rent are also discussed. Other sections explain techniques used to work the several types of pipe used for plumbing. Tips on both working and purchasing are presented to the reader. There is a whole section on typical homeowner plumbing jobs, including what to do when things go wrong in the middle of the night.

The book also covers a variety of installation jobs from minor appliances to major ones, like installing bathroom fixtures. Finally, the work presents a comprehensive view of rural water and waste systems, which are particularly helpful for the owner of a vacation home. No book can be complete without the aid of many people. The Acknowledgments that follow mention some of those who contributed to making this text the most current work possible.

GLENN E. BAKER
REX MILLER
MARK R. MILLER

Acknowledgments

The authors would like to thank the following groups for their help in compiling this book. The first is one of individual people who helped the authors complete the many chores required in preparing a manuscript. These included Bob Baker for helping with some of the photography, and Judy Baker whose expertise kept the computers working despite the proclivities of modern technology. Mary and Jerry at Action Press helped with copies and reproduction. Captain John Baker and Captain Rachel Price-Baker stood tall in the defense of our constitution from all enemies, both foreign and domestic. There were several sources who contributed photographs, illustrations, and technical expertise to this work as well. These included American Olean Tile; American Standard, Inc.; Closet Maid by Clairon; Corl Corporation; Forest Products Laboratory; Formica Corp. (Formica); Gypsum Association (Gypsum); Jacuzzi; Kohler Company (Kohler); NuTone; Owens-Corning Fiberglas Corporation (Owens-Corning); PlumbShop; and, The State of Texas.

1
CHAPTER

Plumbing Systems

RESIDENTIAL BUILDINGS REQUIRE WATER. Water is used for laundry, bathing, drinking, and moving human waste from the house. A modern home uses water in the kitchen for the sink, dishwasher, disposal, and refrigerator ice-maker. Water is needed in bathrooms for toilets, lavatories, bathtubs, and showers. Water is also used for the laundry, outdoor faucets, and lawn sprinklers. Provision must be made for both hot and cold water service.

To provide for many uses for water in the home and for the disposal of human waste, a complex system of pipes is used to carry water to the building and its appliances. Pipes are also needed to carry away waste water.

In some cases, the term *plumbing* includes pipes used to move hot water from furnaces to various rooms of the house for heating. Other plumbing applications involve moving solar heated water into either the hot water system or a central heating system. In other instances, the black pipes used to carry natural gas (or propane in some locales) to a furnace are also considered part of the plumbing system. It is recommended that only steel pipe be used for fuel gas.

The plumbing for a building is usually done in three stages. These are installation of (1) the main outside supply pipe and the main drain; (2) inside pipes, vents, and drains; and (3) water-using fixtures. The first two steps are the focus of this chapter. The third step is usually done after the finished floor is done.

The first step, installing the main supply pipe and main drainpipe, is done at the time of the excavation for the footings and foundation. Pipes for both water and waste are called *lines*. They also might be called the *water line* or the *sewage line*.

Because both water supply lines and drain lines carry water of some sort, both must be located sufficiently underground to prevent freezing during the winter. This is why they are generally put in when the holes and trenches for the footings and foundations are dug. It is cheaper and more efficient to do all the digging at the same time.

The main supply line typically runs from the building to the water meter. See Fig. 1-1. In some areas, the water meter is located in the basement of the building; however, the trend today is to locate the water meter near the street. There it is much easier for the city to read and maintain. The city connects the water main to the meter, and the builder connects the meter to the building. The city will not turn on the water until the plumbing has been inspected and approved.

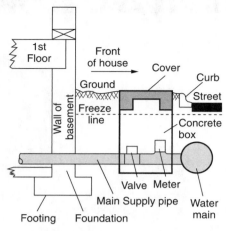

Fig. 1-1 *Initial supply pipes.*

The main supply line and the main drainpipe that carries away waste water must be installed. See Fig. 1-2. Notice that both the supply and drain lines must be buried below the freeze line. Both pipes carry water and are subject to freezing outside the heating building space.

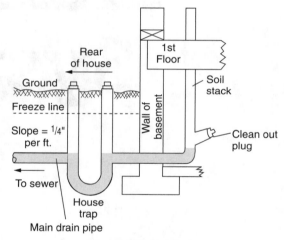

Fig. 1-2 *Initial drains.*

Water and sewage lines are placed into the area that will become the building, and then they are *stubbed out*. See Fig. 1-3A to D. This means that the pipes are located in the areas where they are to be used—the kitchen or bath. Some extra is left sticking up to make full installation a bit easier at a later time. The holes are filled in, and footings, foundations, floors, and such are built over the stubbed pipes.

The second stage of the plumbing is done after the floors and wall frames are started. The remaining pipes, drains, and vents are installed as the building is

Slab

Plastic water pipe in a bed of sand.

(A)

(C)

(B)

(D)

Fig. 1-3 *Supply lines and drains are "stubbed" off for later work. (A) Main line from street to house. (B) Plumbing vents and drains as well as soft copper lines encased in concrete slab. (C) Vents, drains, copper lines in concrete slab. (D) Vents and drains installed before slab is poured.*

erected. Because the pipes and drains are hidden in the walls and under the floors, as in Fig. 1-4, they should be installed before the inside wallboard is applied.

For buildings erected on concrete slabs, much of the plumbing is placed under the slab. The pipes should be laid under the plastic vapor barrier, and there should be some flexibility left in the pipe. This is so because concrete has a high expansion and contraction rate with temperature changes. Leaving some slack in the lines, such as wide curves in copper lines, allows the pipes to move with the expansion and contraction of the slab. Some builders prefer to locate all pipes outside the slab for ease in repairs and service, while others locate the pipes in the attics. Attic installations require careful attention to pipe insulation to prevent the pipes from freezing and bursting in the winter. This can cause serious water damage to the building.

Fig. 1-4 *Plumbing pipes roughed in between inside and outside walls.*

PLUMBING SYSTEMS

A plumbing system consists of supply lines that bring water into the building and drains that carry the water away from the building. For the drains to work efficiently, they must be vented to allow air to enter the system. Thus, the plumbing system has three major parts: a supply, a drain, and vents. The system is shown in Fig. 1-5.

Supply Lines

As mentioned earlier, one water line carries water from the main to the building. Common sizes for the first supply line are 1 inch, 1½ inches, and 2 inches in diameter. Once the pipe is inside the building, smaller pipes are generally used to carry the water to various locations therein. Common sizes for this are 1-inch, ³/₄-inch, and ½-inch diameters.

Water lines can be assembled through holes in floors, wall studs, and so forth. Most water line assemblies are made through openings planned for them. It is a good idea to only run smaller lines through holes cut in wall studs or floor joists. There are two reasons

for this. First, any hole or cut into a joist or stud can weaken it. Second, pipes in these holes might be struck by nails used for hanging wallboard. The nails could cause holes in the pipe that would not be detected until after the wall or floor was sealed and finished and the water turned on.

The supply pipes to most locations are capped with cutoff valves. See Fig. 1-6. This allows the water to be turned off to install the fixture or to make repairs without turning off the water supply for the entire building. Special flexible lines, such as the plastic line in Fig. 1-7, can be used to connect the supply valve to the fixture.

Air chambers must be installed near each supply outlet to prevent a loud noise from occurring each time the water is turned off. The noise is called *water hammer*. When the water is turned off, the full force of the moving water (remember that standard city water pressure is 80 pounds per square inch, which is a lot) slams against the valve. This force can make a loud banging noise, and the force can cause the pipes to physically jump.

The air in the chamber is momentarily compressed to cushion the force of the water stopping against the valve. Figure 1-8 shows this action and the proper location of the air chamber. It is important to remember that each supply outlet must have an air chamber. Air chambers are usually made of the same size pipe as the supply line.

In addition to using air chambers, the pipes should be anchored to something solid, such as a stud or joist, at appropriate intervals. The movement of the water, particularly when it is stopped or started, can cause the pipes to move. The movement can be enough to cause the pipes to "bang" against the floor or wall. Anchoring them reduces both the movement and the noise.

Drains

Drain systems (often called *DWV* for *d*rain, *w*aste, and *v*ent) are used to drain away wastewater. Wastewaters are divided into two categories. The first is called *gray* water and consists of the wastewater from sinks, laundry, and bathing units. The second type, *black* water, contains both liquid and solid wastes from toilets. In most communities, both types of wastewater are drained into one sewer system.

Pipes that carry wastewater from a fixture (such as a sink) are called *drains*. The drains may empty into another larger pipe to be carried across the building to the main drain. These parts that carry the water across the building are called *laterals*. The drainpipes that run vertically are called *stacks,* or *soil stacks.*

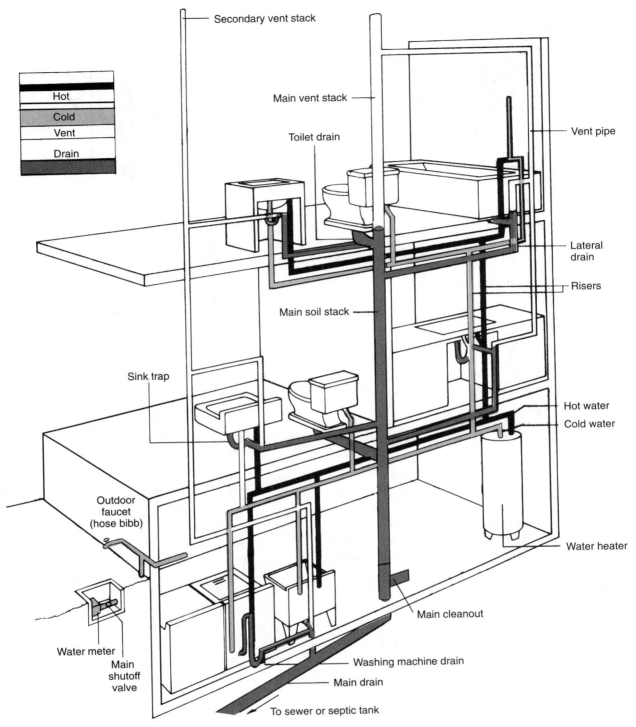

Fig. 1-5 *A typical plumbing system has three parts: (1) supply, (2) drains, and (3) vents.*

Because wastewater is not under pressure, the drain systems must be angled down to allow gravity to move the water. There is typically about ¼ inch of downward slope per foot of horizontal run. This is usually expressed as *3 inches per 12 feet*. This slight slope allows the water to carry the solid waste. If the drain is too steep, the water might run off before the solid wastes are moved and eventually cause clogs, which stop up the drains. Clogs can force the waste to back up into toilets and other drains, preventing their effective use. These backups can have a terrible odor and might spill over onto a floor, causing considerable damage.

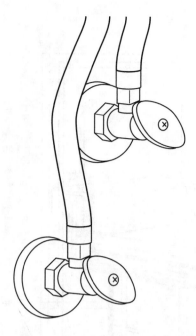

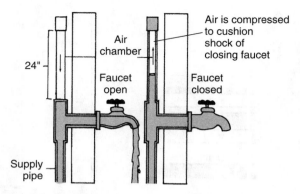

Fig. 1-8 *Typical air chamber action prevents water hammer.*

Fig. 1-6 *Cutoff valves on the pipes that supply fixtures.*

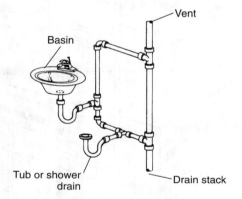

(A). Direct or "wet" vent drains are connected directly to the stack and vent

(B). Reventing (back venting) two drains connected to a stack. Note the loop connecting both units to the vent.

Fig. 1-9 *Typical multiple-drain connections.*

The same slope for the drains must be used from each drain to the main drain line. All drains must slope down.

There is usually only one main drain connecting the building system to the sewer. This means that the several drains in the building must be planned so that they can be connected without eventually causing problems. Figure 1-9A and B shows typical installation factors.

Drains that connect sinks and lavatories, and tubs and showers, to a system are typically 1.25, 1.5, or 2 inches in diameter. Sizes would be determined mainly by available space, building codes, and anticipated drainage flow. The cost of plastic drains is not much different regardless of size.

Laterals can be 2, 3, or 4 inches in diameter. Main drains should be larger in comparison and are typically 4 or 6 inches in diameter. Laterals can be suspended by using *plumber's tape*. This is a perforated metal strip

that allows the support to be installed at a length that will hold the lateral at the correct height to keep the proper drainage slope. See Fig. 1-10.

Another vital part of drains keeps the sewer gases from entering the living space. This is very important

Fig. 1-7 *Reinforced flexible plastic lines make connecting fixtures easy.*

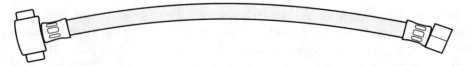

Fig. 1-10 *Plumber's tape holding a lateral drain at the correct slope.*

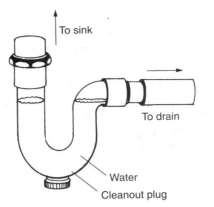

(A). Cross section shows how the water plug prevents gases entering from drain

To sink

To drain

Water

Cleanout plug

because sewer gases are very noxious, can be toxic, and are explosive. A very simple device called at *trap* is used to block out these gases. Older systems might have an *S trap*, but most modern systems use a *P trap*, as in Fig. 1-11. The name is taken because the trap is shaped like the letter P. This keeps a "plug" of standing water between the living space and the sewer. Every time the fixture is used, the water in the trap is replaced, keeping the trap fresh.

Sometimes small objects, such as rings, are accidentally dropped into a sink. The trap will hold them and keep them from tumbling into the main drain, provided that the water is not allowed to continue to flow through the drain.

Traps can be installed with cleanout plugs in them (Fig. 1-11A), or they can be plain, as in Fig. 1-11B. It is not a very difficult job to remove a trap. Both traps allow for the retrieval of lost objects or the removal of solid objects that are impeding the flow of wastewater.

Installing traps is typically not a difficult job. However, a factor called the *critical distance* must be considered. Simply expressed, it means that the outlet into the stack must never be completely below the water level of the trap. If it is, a pipe full of water would also siphon out the water in the trap and expose the living space to the odors and gases of the sewer. It is easy to figure the maximum amount of run you can use on a drain. First, determine the diameter of the drain, such as 1.5 inches. Next, divide by 0.25 inch, which is the amount of slope per foot. This yields the number 6, or 6 feet. However, 6 feet would be the full 1.5 inches, so the maximum run would be anything less than 6 feet. A good rule of thumb is to use only whole numbers for the distance, which would then be 5 feet.

The P traps are used on the drains for every single fixture in the building except water closets. The reason is that water closets have built-in P traps and are usually connected to a fitting called a *closet elbow*.

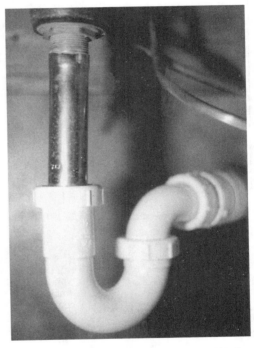

(B). Typical installation without cleanout plug

Fig. 1-11 *Typical P trap installation. (A) Cross section shows how the water plug prevents gases entering from the drain. (B) Typical installation without cleanout plug.*

In many areas, a "house trap" is also required in the main drain outside the building. Check Fig. 1-2. This house trap protects the entire drain system from the odors and gases of the main sewer.

Drain systems are often made entirely of polyvinyl chloride (PVC) plastic pipe of appropriate sizes. However, drainpipes also might be made of copper, cast iron, and heat-resistant glass. Glass drains are fragile and are only used when corrosive materials, such as acids, must be drained from the building to an appropriate collection point.

Vents

As mentioned earlier, vents allow air to enter the drain system so that the water can flow properly. Without the access to air, a vacuum could be formed in parts of the drain system that would prevent the wastewater from flowing. A vent is simply a vertical pipe that rises from a drain up through the roof and allows air to enter the drain, as in Fig. 1-12.

As a rule, a vent is needed for every drain. However, this is often impractical and would require three vents for the typical bathroom, three for a kitchen, one for a laundry, and so forth. The vents can be combined as in Fig. 1-13A and B so that usually there is only one per bathroom, one per kitchen, etc.

Vents are often made from the same type of pipe as the drain system. However, it is common to use the least expensive material for vents because they do not actually carry water.

Clean Out Plugs

Clogs are fairly common in any drain system, and for that reason, special *cleanout* plugs are advisable. These can be located either inside or outside the building, but should be easily accessible. See Fig. 1-14. This allows a cleanout tool called a *snake* or an *auger* to be inserted and used to unclog the drain. If no cleanout plug is used, when a clog occurs, a hole must be dug to reach the drain. Then an opening must be cut into the drain. Needless to say, cleaning out a drain is a dirty job and can be difficult and time-consuming.

It is advisable to install several cleanout plugs. It is even advisable to install a cleanout plug for each drain as well as the main drain. A typical plug is made by

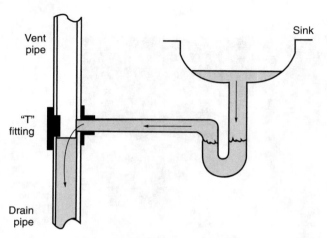

Fig. 1-12 *Basic drain vent.*

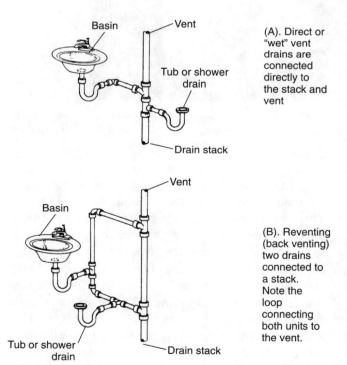

(A). Direct or "wet" vent drains are connected directly to the stack and vent

(B). Reventing (back venting) two drains connected to a stack. Note the loop connecting both units to the vent.

Fig. 1-13 *Multiple vent connections. (A) Direct or "wet" vent drains are connected directly to the stack and vent. (B) Reventing (backventing) two drains connected to a stack. Note the loop connecting both units to the vent.*

Fig. 1-14 *A kitchen cleanout plug located for convenient access.*

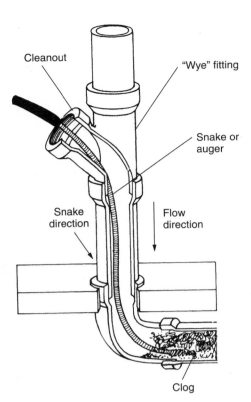

Fig. 1-15 *Cleanout plug detail.*

inserting a "wye" fitting into the line, as in Fig. 1-15. Note that the slanted outlet angles into the drain line to make it easier to insert the cleanout tool. The wye must be the same diameter as the drain and must be assembled to have the least interference with water flow.

2
CHAPTER

Plumbing Codes

WHEN AN INDIVIDUAL FIRST CONSIDERS doing any plumbing, first he or she should become familiar with the building codes for that area. There are plumbing codes for national, state, county, and city (or township) levels. Usually an organization such as a city incorporates all the appropriate codes so that a person doing plumbing in that city has to worry about only one set of codes. When a city or county enacts the law that officially adopts the codes, the law itself is called an *ordinance*. Thus, you may hear or read about either codes or ordinances relating to plumbing.

Most plumbing work fall into one of three categories: (1) repairs, (2) replacement, and (3) new installation. Usually ordinances and codes aren't concerned with routine repairs such as fixing a drippy faucet or unclogging a kitchen drain. However, both replacement of fixtures or pipes and the installation of new work (such as upgrading your old plumbing, building a new house, or adding a new bathroom to an old one) are probably affected by codes.

It is usually easy to find out about the building or plumbing codes in your area. Simply call or visit the city (or county) planning or building permit office and ask for information. They often have summary sheets that give information on codes and the extent of work allowed by amateurs. In some areas, you might even find these data online. Some areas require that all plumbing work be done by licensed and bonded professionals, while others may allow the individual to do the work provided it passes an inspection by a licensed plumber and/or the local building inspectors. Very few areas require no inspection or regulation.

Plumbing codes have been developed over a long time to provide the minimum standards that will protect your health and safety, the public welfare, and property. Codes cover many aspects of the work, including the size and type of pipe used, where multifamily dwellings can be built, where large and small businesses may be located, and many other factors.

For example, suppose that a person owned a very large, old five-bedroom house and decided to partition it into several apartments. If the water and sewage mains had not been designed for multifamily use, they would probably be too small to provide enough water or to carry away all the wastewater. The new apartments could very well be plagued by a poor water supply and clogged sewer lines. Compliance to good building/plumbing codes helps prevent such costly mistakes.

PLUMBING CODES

Most states or provinces, counties (or parishes), and cities have adopted plumbing codes based on one of several national codes. These include the Building Officials Conference of America (BOCA), Uniform Plumbing Code (UPC), Standard Plumbing Code (SPC), or the National Plumbing Code (NPC) in Canada. All the codes cover such factors as

Plumbing design

New installation

Replacement or alteration

Potable (drinkable) water systems

Hot water systems

Waste and sewage disposal

Storm drains

Gas pipes

Each of these codes (and there are some others) provides information on the size of pipe for the required distance and the diameter of drains for sinks, showers, clothes washers, and other fixtures. The codes also specify types of pipe, trap requirements, and sometimes even how far away from living areas baths or toilets must be. Codes are usually updated every 3 years. Because of this updating, it is fairly common to find an installation that met the codes when installed, but will not meet the codes if it is just replaced.

Another element in building codes is the practice of zoning. An area or subdivision is usually *zoned* as one of the following:

Residential

Single-family

Duplex

Multifamily or apartments

Commercial

Industrial

Agricultural

The city or county plans for water, sewer, and drainage systems based on how the land is used. You can usually build a single-family residence in a multifamily zone, or even in an agricultural zone, but you can't build an industry or an apartment in a single-family residential area. The reason is that the water systems built by the city or county aren't designed to handle the larger traffic.

PERMITS

As mentioned before, you probably don't need a permit to fix a leaky faucet, unclog a drain, or even replace worn bathroom sinks and faucets. In these cases, you are not changing the basic location, supply lines, or drains. However, permits are generally required if you build new, enlarge, change size or location, improve flow and pressure, convert from one system to another, or demolish an existing facility.

The first step is to contact your local planning or permit office for information. Next, you must develop a detailed plan of what you wish to do. Most planning offices require a basic job description, an estimated cost, and some sort of diagram. Then you pay a fee based on the estimated cost. The smaller the job and cost, the lower the fee.

You may request information on codes, practices, licensed plumbers, and so forth from the planning or permit office. Meanwhile, you wait. It may take anything from a few minutes to a couple of weeks, depending on the type and complexity of the job and the completeness of your plans.

Once a permit is issued, as in Fig. 2-1, you normally post it so that it is visible from the street. Some jobs will require that various stages of the project be inspected before you can move to the next. The inspection tag in Fig. 2-2 shows that the rough-in plumbing has been approved and that the builder may move on to the next stage, which is to install the various connections and cutoff valves for the fixtures and then to enclose the walls, as in Fig. 2-3. At this point, stubs are also cut off, and mounting brackets are installed for toilets and other fixtures, as in Fig. 2-4.

Fig. 2-2 *A tag is placed on the work when the first stage of the plumbing is approved.*

Fig. 2-1 *A building permit must be visible from the street.*

The inspectors (they may be building inspectors or plumbing inspectors depending on the job and the area you're in) usually have some latitude (wiggle room) in making their inspection. They are most unlikely to approve anything in clear violation of the codes, but they may use some discretion in those matters not covered by the codes. For example, many codes assume that a house will be plumbed using the same type of pipe (such as copper). However, some older homes were originally plumbed with galvanized pipe, but as these pipes rust, an owner may wish to replace sections with plastic or copper pipes. Thus, an inspector would likely approve the different replacements provided the newer materials met all the appropriate codes for size, length, and so forth. Some codes do allow for this type of pipe mixture.

Usually, the materials sold by the established building supply vendors in your area are those that can be easily approved. Most vendors will gladly offer information on what should be used, and they can often provide information on passing inspections.

The inspection must be passed in order to have the water or other utility turned on for use. If an inspection

Fig. 2-3 *Walls are enclosed after inspection.*

Fig. 2-4 *Stubs are cut off and mounting brackets can be installed. This shows the mounting bracket for a toilet.*

is not passed and there are no clear violations of a code, most communities have an appeal process. The best way to do something, however, is to do it well within the codes.

CHECKING YOUR SYSTEM

There is an old saying, "Don't fix it if it ain't broke." However, there are some things you can observe that should be considered fixable. The signs include

Visible water in an area

Drips

Stains or mineral deposits on pipes

Lack of pressure

Drops in pressure or supply when a fixture is used

Knocks, rattles, and squeals

Odors

If you find a visible puddle, water is obviously leaking out of somewhere. Puddles are signs of faulty water heaters, holes or splits in pipes (often behind walls), or loose joints in pipes. These are all things that are fixable and that should be fixed.

Drips are most common at pipe joints and faucets. Joints can be tightened or doped (more on that later), and faucets usually drip because they need new washers or cartridges. Again, more on that later.

Stains or mineral deposits on pipes, as in Fig. 2-5, indicate water seepage. The water flow may be too small to be visible, but it's there. If the area is damp to the touch, it should probably be fixed. If no moisture can be detected, one should usually leave it alone.

Low water pressure and low flow may be caused by several things. First, you may not have enough pressure at your source. There isn't much you can do about that except report it to the authorities. If you have adequate pressure at your source, then the low pressure is caused either by pipes that are too small or by a clog or obstruction in the pipes.

Small pipes either from your source to your house or within your house are common problems. Often, an older house is plumbed with small pipe that was adequate at the time for the needs of the house. Years later, new rooms are added and plumbed with the same size pipe. Sometimes this can be fixed with simply a larger supply line. When it can't be fixed this way, you may want to simply live with it rather than completely re-plumb your house with larger pipe. That can be rather expensive.

There is one other cause of low pressure in an area when the supply pressure is adequate—an obstruction to water flow in the pipe. One of the most common

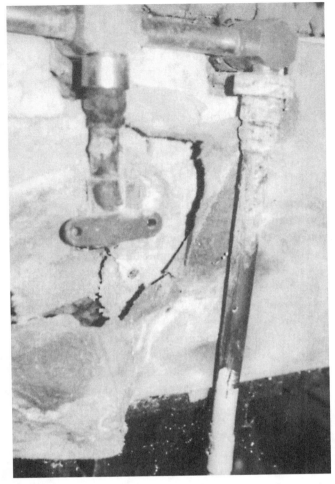

Fig. 2-5 *Mineral stains at pipe joints and crumbling plaster point to seepage.*

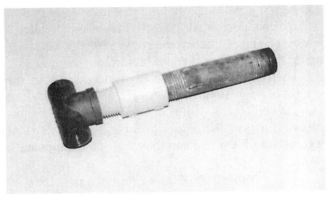

Fig. 2-6 *A plastic fitting between copper and steel pipes will prevent a clog from electrolytic buildup.*

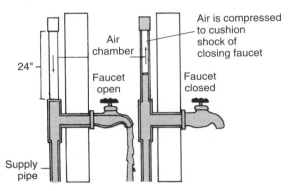

Fig. 2-7 *An air chamber prevents banging noises called water hammer.*

causes of this problem is electrolytic action between two different types of metallic pipe. If you have galvanized pipe joined to copper pipe, you will get an electrolytic reaction between the copper and iron with water as the electrolyte. Over time the resulting buildup will completely block the flow of the water. The solution is simple. You simply separate the copper pipe from the galvanized pipe with a plastic fitting such as a bushing or transition joint. See Fig. 2-6.

Another pressure problem caused by small pipe size is often felt in the shower. Suddenly the water turns too hot because someone in another part of the house has flushed a toilet or turned on a cold-water tap. To fix this, larger supply lines must be installed.

Knocks, rattles, and squeals indicate various problems. Knocks or loud bangs are usually caused by the lack of an air chamber, or one that is full of water rather than air. An air chamber is a short length of pipe near a faucet, as in Fig. 2- 7. These can be installed in out-of-the-way places instead of tearing out walls. The banging noise is caused when the water is abruptly

turned off and slams against the closed tap. The air chamber simply cushions the surging water when it is turned off. Rattling pipes are usually loose. Pipes should be supported or fastened in place at regular intervals. The exact distance between intervals also depends on the size and type of pipe. These distances are given later in the data about standards. Squeals are usually caused by washers that are loose or damaged. Washers are fairly easy to replace.

Finally, another indication of a plumbing problem is the presence of a foul odor. The most common causes of odors are a problem in the vent system or a bad wax ring under a toilet. A sniff test to locate the source of the odor will usually pinpoint the cause.

GENERAL CODE STANDARDS

The following tables give a composite view of the requirements to meet most standards. Plumbing codes focus on the three elements of supply, drains, and vents. All three have standards concerning size of pipe, types of pipe, locations, and supports. To interpret the codes, first you must determine your unit requirements. Units are determined by function, such as washing clothes and taking a shower. Anything that uses water is usually attached, or "fixed," to the building and is called a *fixture*. Fixtures that use more water carry a heavier unit rating than ones that use less water.

There is some variance in the codes about the weight of a function, such as taking a shower. The best thing to do is check with your local planning or permits office. However, the figures in Table 2-1 will give you a good idea of how to determine your unit requirements.

Table 2-1 Fixture Unit Ratings

Fixture	Unit Rating
Bar sink	1
Bathtub	2
Clothes washer	2
Dishwasher	3
Kitchen sink	2
Shower	2
Toilet	4
Utility sink	2
Vanity sink (lavatory)	1

To give an example, let's count the units for a three-bedroom house with two baths and a laundry area. Assume there is no basement, and the laundry area has only a hookup for a washer and a dryer. Also assume that both baths have a combination shower/tub. The kitchen has a sink and a dishwasher.

Kitchen sink	2
Dishwasher	3
Clothes washer	2
2 bathroom (vanity) sinks (1 ea.)	2
2 toilets (4 each)	8
2 shower/tubs (2 each)	4
Total	21

From this total, you can now use the following tables that relate to pipe size, drain size, and so forth.

Supply System Codes

The size of the pipe to the home and then to the various places where water is used is important. If the pipe is too small, then low water flow and dropping pressures will be experienced. Once you know the number of units in your home, you can determine the size of pipe needed. Two tables are given. Table 2-2 is for low water pressures, typical of many rural, electric pump/pressure tank systems. These systems usually cut on when the water pressure in the tank drops to about 30 pounds per square inch (psi) and cut off when the pressure reaches 50 or 55 psi. The second table, Table 2-3, is for higher pressures typical of most cities. Remember that the nominal standard city water pressure is 80 psi. Most water flow engineering tables use that pressure. However, there is a lot of variance. For that reason the data in Table 2-3 are a composite table that should cover about anything over 60 psi. If you don't know what you have in your area, ask when you call the local office.

If the house in our example, with 21 units, is in a rural location with low pressure, you will note that no

Table 2-2 Water Supply Lines for Systems under 60 psi

Size from Water Main to Meter, Inches	Size from Meter to Home, Inches	Length of Pipe Allowed, Feet						
		40	60	80	100	150	200	400
		Fixture Units						
3/4	3/4	20	21	19	17	14	11	6
3/4	1	39	39	36	33	28	23	21
1	1	39	39	39	36	30	25	18
1	1 1/4	78	78	78	78	66	52	33

Table 2-3 Water Supply Lines for Systems over 65 psi

Size from Water Main to Meter, Inches	Size from Meter to Home, Inches	Length of Pipe Allowed, Feet						
		40	60	80	100	150	200	400
		Fixture Units						
3/4	3/4	21	21	20	20	17	13	8
3/4	1	39	39	39	39	35	30	27
1	1	39	39	39	39	38	32	22
1	1 1/4	78	78	78	78	74	62	39

more than 60 feet of pipe is allowed if the pipe used throughout is ³/₄-inch pipe. However, if you switched to 1-inch pipe in the building, even though the supply line remains a ³/₄-inch line, you could have up to 250 feet of distribution pipe in the building.

If the house in our example is in a city with pressures of more than 60 psi, then we could have about 80 feet of distribution pipe. If we increased the size of pipe in the house to 1-inch pipe, then we could have 400 feet of distribution pipe in the house.

Each of these figures is actually a fairly generous amount. Most modern homes are planned to concentrate water-using areas into two or three main areas. That's why kitchens and laundries are often close, while bathrooms may be located near each other. But, as you can see from the tables, by slightly increasing the size of a supply line, you can greatly increase the length of the lines you can use. This again shows that codes are minimum standards rather than ideal ones. If you are building new stuff, it's a good idea to go one step beyond the codes. It makes it easier for the inspectors to approve, provides better service in the long run, and keeps you in compliance with minimum codes for a longer period.

Drain System Codes

Drain system design is just as important as the design of supply systems. The same unit values are used to determine the size of the drain needed to carry away the used water. Drain system design is rated in *trap size*. It's really about the same as citing the diameter of the drain, and it is all based on how much water the drain will carry away in a given time.

Chapter 1 showed two types of traps, an S trap and a P trap. See Fig. 2-8A and B. The S traps (Fig. 2-8C) are no longer approved by most codes, but you are allowed to replace an S trap in an older system. However, they are not as safe and reliable as the newer P traps.

Every drain should have a P-type trap, as in Fig. 2-8, installed with it. The only drain that does not require a P trap is a toilet. That's true because all approved toilets have P traps built into them. See Fig. 2-9. While we are on the subject of toilets, the official architectural name for a toilet is a *water closet*. It is often shown on plans with a *WC* designation. In some places, it's also called a *commode*. It was named *water closet* by its inventor, an Englishman named John Crapper. In England, they call it a *loo*.

The code also sets the distance that a drain can be from a vent. Because drains are powered by gravity and not water pressure, air must be allowed into the drain for the water to drain out. If a drain fills with water

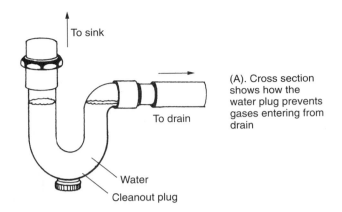

(A). Cross section shows how the water plug prevents gases entering from drain

To sink

To drain

Water

Cleanout plug

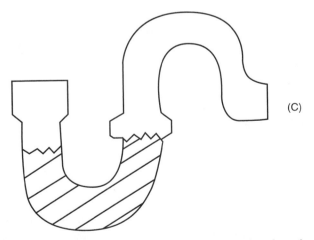

(B). Typical installation without cleanout plug

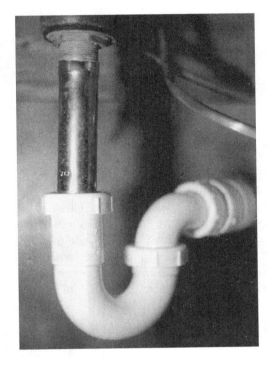

(C)

Fig. 2-8 *Typical P trap installation. (A) Cross section shows how the water plug prevents gases from the drain. (B) Typical installation without cleanout plug. (C) The S traps are no longer approved for new construction but may be replaced.*

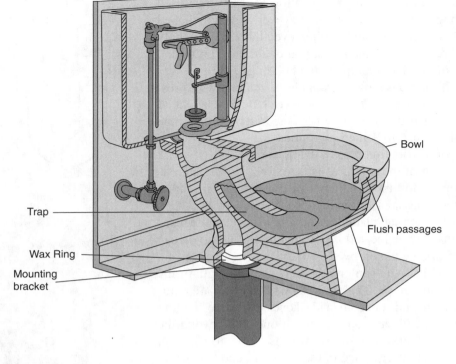

Fig. 2-9 *Toilets have built-in traps.*

Bowl

Flush passages

Trap

Wax Ring

Mounting
bracket

too quickly, this will trap air in the drain. Trapped air, without a vent, can cause the drain to siphon out the trap and open the house to gas and odors. This makes the proximity of a vent to every drain a real necessity. The distances indicated in Table 2-4 provide data on locating vents in drains of any given size. Remember that the size of the drain is determined by the unit rating of the fixture. These data are also given in Table 2-4.

Table 2-4 *Trap Size Requirements and Distances to Vents*

Fixture	Unit Rating	Minimum Trap Size, Inches	Maximum Distance from Trap to Vent, Feet
Bar sink	1	1¼	2.5
Bathtub	2	1½	3.5
Clothes washer	2	2	5
Dishwasher	3	1½	3.5
Kitchen sink	2	1½	3.5
Shower	2	2	5
Toilet	4	3	6
Utility sink	2	1½	3.5
Vanity sink (lavatory)	1	1¼	2.5

The code also specifies the unit capacity of drains. Look at Table 2-5. A typical kitchen sink is equipped with a drain and trap 1½ inches in diameter. A kitchen sink has a unit rating of 2. Thus, the drain and trap can handle the outflow of the sink if it is a vertical drain. However, if the drain must run horizontally for more

Table 2-5 *Drainpipe Unit Capacities*

Drain Size, Inches	Vertical Drain Units	Horizontal Drain Units
1¼	1	1
1½	2	1
2	16	8
2½	32	14
3	48	34
4	255	215

than a few inches, then the 1½-inch diameter is not enough. If you look at the table, you will notice a dramatic increase in capacity from 2 to 16 units if you simply increase the size of the drain from 1½ to 2 inches.

What is commonly done to take advantage of this increased capacity is to connect several fixtures, such as a kitchen sink and a clothes washer, to a single 2-inch vertical drain. This is also a good idea from a practical point of view. For example, most dishwashers are designed to discharge through the kitchen sink system. Most kitchen sinks use drains that are 1½ inches in diameter. To prevent backups, the kitchen drain could empty into a larger drain within a short distance. From Table 2-5, you can see that the capacity increases from 2 units to 16 units.

Most codes now require the installation of a house trap, as in Fig. 2-10. The house trap blocks odors and gases from the sewer mains. This is important because sewer gases are often poisonous and explosive. They don't smell good either.

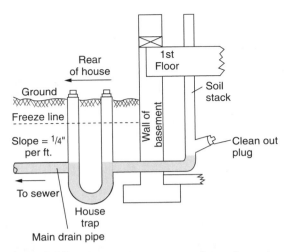

Fig. 2-10 *House traps are newer code requirements.*

Vent Systems

The size and location of vents are also governed by the codes. Two factors are critical—size and distance from the fixture served. Remember that drain systems are gravity-powered and must have vents to allow the water to flow. Vent pipes are usually a bit smaller than drains because air is less dense than water. From Table 2-6, you can see that the 2-inch drain in our previous illustration requires only a 1½-inch vent. Vents serve another important function. They exhaust gases and odors from the system in your house into the air. Most codes call the gases noxious and dangerous. Just as sewer gases can, they smell bad and can explode. Vents usually run up through the roof, as in Fig. 2-11. These gases and odors are usually lighter than air, and venting them at roof level allows them to rise and dissipate without being in the human living area.

Table 2-6 *Code Requirements for Vent Sizes and Locations*

Fixture Drain Size, Inches	Minimum Vent Sizes, Inches	Maximum Distance from Trap, Feet
1¼	1¼	2.5
1½	1¼	3.5
2	1½	5
3	2	6
4	3	10

Support Codes

Another code requirement is to support the pipe used in all three systems—supply, drain, and vent. Table 2-7 shows the support required for different types of pipe. As

Fig. 2-11 *Drains are vented with pipes that go up through the roof.*

Table 2-7 *Pipe Support Codes*

Pipe Type	Horizontal Intervals, Feet	Vertical Intervals, Feet
PVC	3	3
CPVC	3	3
ABS	4	4
Rigid copper	6	10
Flexible copper	4	8
Steel		
⅛ to ¾-inch diameter	10	15
Over ¾-inch diameter	12	15
Cast iron	5	15

a rule, it doesn't matter what size the pipe is or how many units it serves. Lack of proper support can cause pipes to rattle or sag. Sagging can sometimes cause joints to break and leak.

There are several different types of support methods. Some depend upon spacing, location near a solid framework, or other factors. Codes often do not specify the type of support, only the spacing of the support. Building inspectors can blow the whistle on anything they think is inadequate.

Typical supports include plumber's tape (metal strips perforated at intervals) and a variety of brackets, as in Figs. 2-12A to C, 2-13, and 2-14.

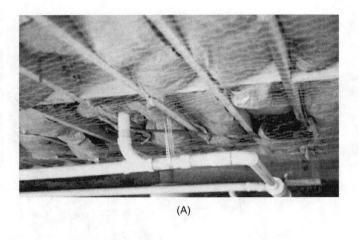

(A)

Plumber's
tape

(B)

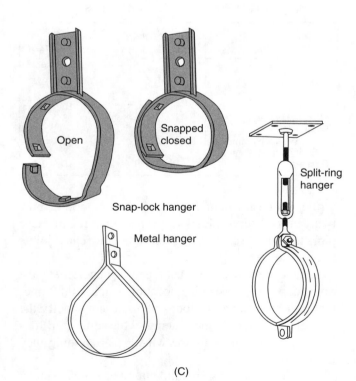

Open

Snapped
closed

Snap-lock hanger

Metal hanger

Split-ring
hanger

(C)

Fig. 2-12 *(A) Plumber's tape holding a lateral drain at the correct slope. (B) Plumber's tape is sold in rolls. It is nailed in place. (C) Other types of hangers.*

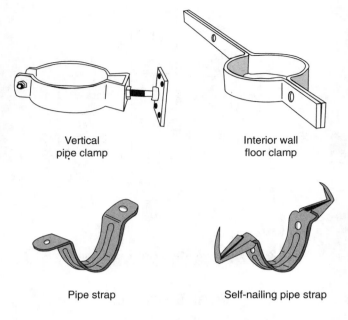

Vertical
pipe clamp

Interior wall
floor clamp

Pipe strap

Self-nailing pipe strap

Wire strap

Self-nailing
wire strap

Fig. 2-13 *Various types of straps are used to support both horizontal and vertical pipes.*

Fig. 2-14 *A fitting for copper pipe with an integral brace. These are "drop" fittings.*

Pipe Type Codes

Pipe type is also governed by codes. Factors considered in the codes include durability, sensitivity to heat or corrosion, strength, resistance to soil shifting, and so on. Each type of pipe has good points and bad points. Some types of plastic pipe can't be used for hot water because the heat softens both the pipe and the joints. Pipe type is critical to the efficiency and long life of a system. Table 2-8 gives a composite view of the many factors in

Table 2-8 *Pipe Type Comparisons*

Pipe Type	Color	Cost	Work Difficulty	Weight	Corrosion Resistance
Copper pipe	Red	5	4	5	8
Copper tubing	Red	6	5	5	8
PVC	White	2	2	2	10
CPVC	Cream	2	2	2	10
ABS	Black	1	4	2	10
Galvanized	Silver	7	8	7	4
Cast iron	Black	8	10	10	8
Clay	Red	6	8	10	10
Fiber	Black or aluminum	4	6	4	7

Note: Factors are scaled from 1 to 10, with 10 = higher number.

pipe selection. Some codes specify types of pipe that can be used on the basis of service life, resistance to freezing, or soil shifts.

The strongest type of supply pipe is steel, either galvanized or black pipe. However, it rusts. Black (uncoated) pipe rusts quickly while galvanized pipe may last 20 or more years. One of the authors ran a galvanized supply line to a house in 1964 that is still in use. Plastic, on the other hand, doesn't rust, is easy to work, and is comparatively cheap. It could theoretically last almost forever. However, it isn't very strong and can deform from heat. Knowledge of these factors can help you make a good choice.

Another factor in some codes is what type of pipe can be used for what function, such as supply, drains, or vents. Builders are always looking for a less expensive material, or one that can be worked faster. There are many types of pipe that can be used for one function, but not another. Table 2-9 outlines some typical standards.

Some codes also specify what must be done when both the supply main and the drain or sewer are located in the same trench. In all our previous illustrations, the supply line was located in the front of the home while the drain led to the rear.

In many places, both systems are located in the same trench. When this is done, the supply line is always located above the drain. This minimizes possible pollution if a drain begins to leak. Figure 2-15 illustrates

this. This rule also applies to pipes that cross as well as those that run parallel.

Codes usually specify that cleanout plugs must be installed and where they are located. In southern states where basements are less common, the plugs are typically located outside the home. Where basements are more common, cleanout plugs are commonly required in the basement. See Figs. 2-16 and 2-17. Codes may

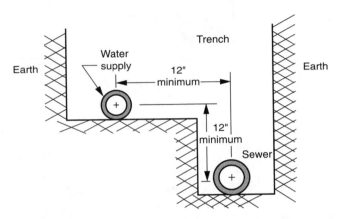

Fig. 2-15 *Code requirements when a supply line and a sewer line are in the same trench.*

Table 2-9 *Common Pipe Type Applications Allowed by Codes*

Cold supply lines	Copper, galvanized, PVC, CPVC
Hot water lines	Copper, galvanized, CPVC
Drains	ABS, PVC-DWV,* cast iron, copper, galvanized†
Vents	Same as for drains
Sewers	Cast iron, PVC-DWV, ABS-DWV, vitrified clay, fiber

*PVC-DWV can only be used for up to three stories because of its high expansion and contraction rate.
†Galvanized drainpipe must be used aboveground. It should not be buried.

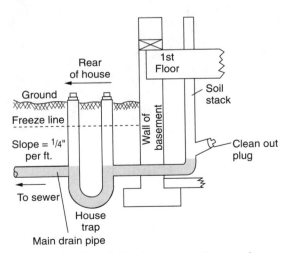

Fig. 2-16 *A basement cleanout plug.*

Fig. 2-17 *An external cleanout plug.*

also specify the location of water meters. In states where the soil does not deeply freeze in the winter, meters are located near either a street or an alley. They are enclosed in "boxes" buried a foot or so in the ground, as in Fig. 2-18. This allows the city meter readers quick and easy access to read the meter. In colder climates, meters are often located within basements.

Finally, national codes are usually zoned. Fig. 2-19 shows the zones within the continental United States. Codes for the various zones are developed to allow for variances in building practices and state regulations. Builders within these zones become familiar with the factors in that zone, and so building practices vary a good bit from one part of the nation to another. National codes are guidelines provided by appropriate professional associations. They are not laws. When a city or county formally enacts an ordinance that requires builders to adhere to a code, that is when that code becomes a law. A very large number of cities and counties use only parts of the codes as guidelines and then establish their own guidelines appropriate for their location. City and county ordinances are laws and take precedence over any national guideline.

Fig. 2-18 *A meter box installation.*

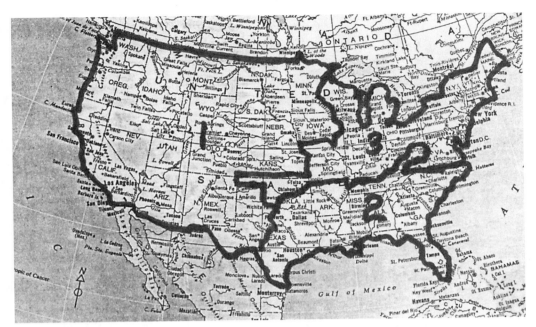

Fig. 2-19 *Approximate boundaries of national plumbing code zones.*

3
CHAPTER

Before You Start

O FAR, WE HAVE COVERED SOME BASIC IDEAS and a good deal of technical information about codes. There is still a lot to consider before undertaking a major plumbing job. Probably, the first thing to check is whether you need to do any detailed planning. So many of us just decide to start something and then adjust as we progress. Some say, "Play it by ear." Plumbing can be hard on ears. We made several more trips to the hardware store or building supply center than we really needed to if we had stopped to think a bit. Ah yes, we've been there, done that, and mopped up after.

Earlier we said that plumbing jobs usually fall into one of three categories: (1) repair, (2) replace, and (3) new construction. Simple repairs such as fixing a leak in a pipe, a drip in a faucet, or a flush valve in a toilet probably don't need plans or permits. There are simple replacement jobs, too, that require little, if any, formal planning. Examples are changing a worn, mineral-crusted faucet on a lavatory, or replacing a chipped kitchen sink. Even on these, however, at least some mental planning is advisable.

Some of us have proceeded without investigating at some time, and we have discovered that the "simple" project ended up inside a wall with more complications than we imagined. There were also the wails of family members because the water was turned off a lot longer than anticipated.

STEPS IN PLANNING

Probably the first thing to do is to check out your present system very carefully and then decide exactly what you want or need to do. Literally, decide exactly what you want as an end product, and then determine what you need to do to get to that desired end from what you already have in place. Then you can select the appropriate type of pipe, fixtures, cabinetry, and tools you need. Once you know what is to be done, you can make estimates on time, cost, and so forth.

CHECK OUT YOUR SYSTEM

To check out your system, you need to look at everything. Find out where your water supply line enters your home. Find out where the meter is and how to turn off the water to your whole house. Make a rough sketch of this, as in Fig. 3-1.

Then follow the supply lines as they run through the house. You should record the size of the pipes, type of pipe used, any problem signs that pop up, and any valves that are in place. In some areas, homes may have several valves that cut off the water supply from different areas without disturbing the flow in the rest

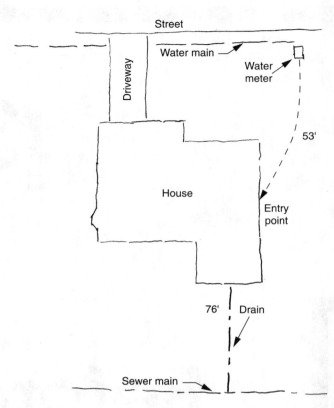

Fig. 3-1 *Locate water supply, meter, and drain in a plot plan.*

of the house. Include the distances of the runs and the distance between supports, valves, and feeder lines. Do this for both cold and hot water systems. Also, do this for every floor of the house, including the basement. On your notes, be sure to include information about the locations of pipes. Are they between walls? Are they under a ground floor, but visible from a basement or crawl space? Then make a sketch of this information, as in Fig. 3-2.

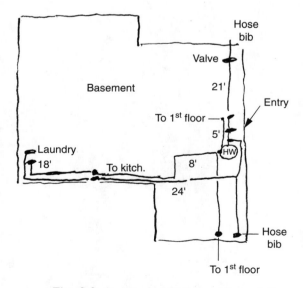

Fig. 3-2 *Sketch a supply diagram for each floor.*

Next, you should do the same with the drain, waste, and vent (DWV) system. Again, make a rough sketch of this information, as in Fig. 3-3. Be sure to include sizes and distances. If you can't find out exactly where a pipe is, indicate a general area with a notation or question mark to show that the location is not exact.

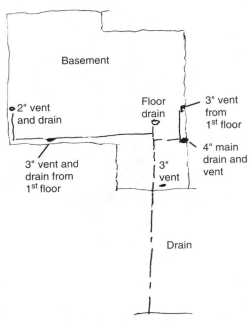

Fig. 3-3 *Sketch your DWV system.*

To help you in making a decent drawing (but keep your notes and rough sketches just in case), Fig. 3-4 shows the symbols used in pipe drawings. You can use these to add to your skills and knowledge, or use them for reference as needed.

All this information is useful if future projects are started as well. This information may be required to obtain a building permit for substantial improvements or additions. For major projects in some areas, it will be necessary to convert your data into very neat and precise working drawings that must be submitted to the planning and permit office. Sometimes, working drawings may be done by almost anyone and submitted on almost any size or type of paper. In strict areas, drawings must be done by professional drafters and must be on specific sizes and types of drafting films.

All this information, then, can be sorted into three groups: rough notes, sketches, and working drawings. To find out what level of expertise is required in your planning, again, check with your planning office.

PIPE SELECTION

There are a number of things to consider when you select the materials you want to use. First, and foremost,

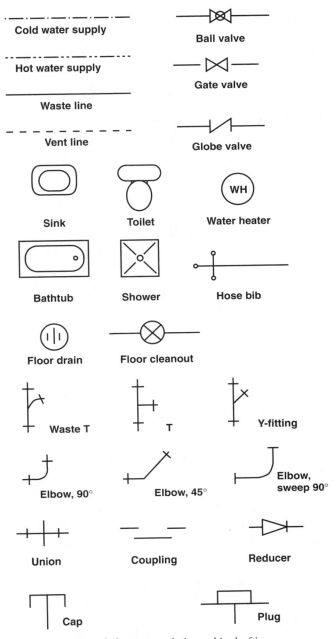

Fig. 3-4 *Pipe symbols used in drafting.*

you should select the type and size of pipes, fixtures, and supports that are within the building codes for your area. Of course, cost is usually a factor as well as appearance, durability, and ease with which the job can be done. In most cases, you can replace one type of pipe, such as galvanized, with a different type, such as plastic. However, you need to know how to make the transition.

There are numerous types of pipe. Some can be used for almost any purpose while other types can be used only for drains and sewers. Some discussion of this was given in Chap. 2 on codes. The types of pipes generally fall into three groups: (1) metallic, (2) plastic, and (3) clay and fiber. You might also group them into supply lines, or drain, waste, and vent (DWV) pipes.

When you make a turn, join two or more pieces together, or go from one size to another, pieces called *fittings* are generally used. Figure 3-5 shows a number of the types of fittings that are used. They are used in working most types of pipe and are available in iron, steel, copper, and plastic for iron, steel, copper, and plastic pipes, respectively.

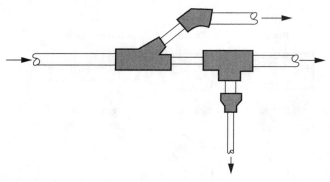

Fig. 3-5 *Fittings let you work pipe to change size and direction.*

Metallic Pipe

Metallic pipes are perhaps the oldest type used. Lead pipes were used in Roman times, and cast iron pipes have been in use for over a century. Time and experience have provided some insights into the best uses to avoid such outcomes as lead poisoning and so on. Metallic pipes include pipes made from cast iron, steel, lead, brass, and copper.

Cast iron pipe Cast iron pipes are made of real cast iron. They are tough, durable, strong, and heavy. They are used almost exclusively for main DWV pipes and are not available in smaller sizes appropriate for supply lines. They are difficult to cut and complicated to seal. Many codes now require repairs and replacements to be made with plastic pipe. Table 3-1 shows information on metallic pipes.

Steel pipe Steel pipes are often called iron pipes. Iron is, of course, the main ingredient of steel, but there is a difference. Almost all threaded pipe is steel pipe. Steel falls into two broad groups: galvanized pipe and black pipe. Both are joined and direction is changed with pieces called *fittings*.

Galvanized pipe As in Fig. 3-6, galvanized pipe has been coated with zinc inside and out to reduce corrosion. This reduces corrosion; it doesn't stop it. Galvanized pipe may last a very long time, but if you have galvanized supply lines and notice rust stains or a rusty taste to the water, it's because the corrosion is advanced.

Fig. 3-6 *Galvanized pipe and fittings.*

Black pipe Black pipe doesn't have a coating. It's called black pipe because a black scale is formed on the surface when it is shaped at the steel mill. Sometimes the steel mill will apply a thin coat of varnish or lacquer to the pipe. See Fig. 3-7. This type of pipe is

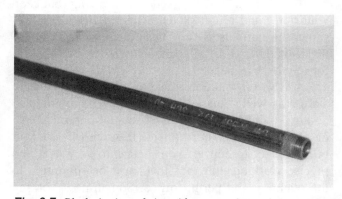

Fig. 3-7 *Black pipe is steel pipe with no corrosion-resistant coating.*

Table 3-1 *Metallic Pipe Factors*

Type	Use	Diameters, Inches	Std. Lengths, Feet	Joining
Rigid copper	Hot and cold water supply	$3/8$, $1/2$, $3/4$, 1	10, 20	Soldered fittings
Flexible copper	Hot and cold water supply and gas lines	$1/4$, $3/8$, $1/2$, $3/4$, 1	30 or 60 coil, 10 and 25 coils in some areas	Flare or compression fittings
Galvanized steel	Hot and cold water supply and DWV	$1/8$, $1/4$, $3/8$, $1/2$, $3/4$, 1, $1 1/2$, 2	21 "joint"	Threaded fittings
Brass	Valves, special drains	$1/4$ to $1 1/2$	Varies—specialty items	Special fittings
Cast iron	Main drains, vents	3, 4	5, 10	Packed and leaded fittings

normally used only for gas lines in homes. It will rust out too quickly to be used for water or chemicals. Galvanized pipe may also be used for gas lines, but is a bit more expensive.

Copper pipe Copper pipe is perhaps the most commonly used pipe for supply lines. See Fig. 3-8. There are two types of copper pipe, *rigid* (or *hard*) and *flexible* (or *soft*) copper. The rigid copper is usually called *pipe* while the flexible copper is accurately called *tubing*. Both are relatively light and inexpensive, but both require some special tools and skills. Also, there are three classifications, or grades, of copper pipes based on the wall thickness of the pipe. Type M has the thinnest wall and is probably the easiest to work. The thickest is type K, with type L being somewhere between thick and thin. Flexible copper, however, has only two grades, K and L. Table 3-2 compares the grades for rigid and flexible copper.

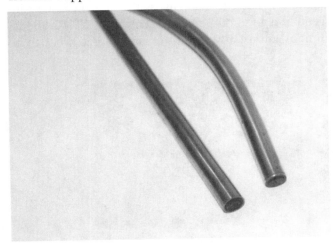

Fig. 3-8 *Rigid copper pipe with soft copper tubing above it. Note the slight curve in the soft copper from being coiled.*

Table 3-2 *Copper Pipe Grades and Fitting Process*

Wall Thickness Grade		Rigid Copper	Flexible Copper
Thin	M	Solder or compression	—
Medium	L	Solder	Flare or compression
Heavy	K	Solder	Flare or compression

Note: Flexible copper may be fitted by soldering, but special tools may be required.

Another point on working with copper is that it takes special tools to join rigid pipe to flexible copper tubing. The fittings for rigid pipe will not match or join with those for flexible tubing. One must use either special fittings or tools called *swages* to enlarge the end of the soft copper tubing to be like a built-in fitting for the

rigid pipe. This will be mentioned again later, but this kind of information is helpful for planning.

Some copper pipe and tubing are plated with chromium. This makes its appearance more attractive and gives protection from stains and corrosion. Such chrome-plated copper is often found on visible water lines and drains under open sinks.

Lead pipe Lead pipe is not used very often anymore. Building codes allow it to be used for DWV, and lead sleeves are still used on roofs to provide weather seals around vents. See Fig. 3-9. However, it is very heavy, soft, and rather expensive. It should never be used for a supply line.

Fig. 3-9 *A lead vent sleeve has been worked over the vent to seal the hole made for the vent.*

Brass pipe Brass pipe is commonly used for traps, drains, and connecting lines. It is one of the most expensive types of pipe. For household uses it is usually chrome-plated for appearance. Brass pipe is heavy, difficult to work, and expensive. Because copper is the main ingredient of brass, you can use copper and brass fittings interchangeably. You may also come across an occasional bronze fitting or valve. Brass is an alloy of copper and zinc while bronze is an alloy of copper and tin. Other metals, such as antimony, may be used in the alloys for various reasons. For amateur plumbing, there probably is little difference between the two. Brass will corrode more than bronze, which is why many valves are made of bronze rather than brass.

PLASTIC PIPE

Plastic pipe, as in Fig. 3-10, is rapidly becoming the choice of both amateur and professional plumbers. It is inexpensive, quick and easy to work with, tough and durable, and very reliable. Plastic is also a better insulator than metal, and this makes it somewhat more resistant to freezing and easier to protect. It does have drawbacks such as heat deformation, a high rate of expansion and contraction, softness, and relatively low strength. However, if you adjust for these factors, plastic pipe is a good choice.

There are several types of plastic pipe, as depicted in Table 3-3. These include polyvinyl chloride (PVC), chlorinated polyvinyl chloride (CPVC), acrylonitrile butadiene styrene (ABS), polyethylene (PE), and polybutylene (PB). Some are rigid, and others are flexible. As with copper, there are different fittings for rigid and flexible plastic pipe. Almost all the various plastics are very versatile and can be used for supply lines as well as for DWV. There are some limitations, and local codes can highlight these.

Plastics, on the whole, do not corrode or stain from minerals. White plastic drains and supply lines are commonly used on open sinks where they are easily visible. They retain a good appearance for a very long time.

PVC can be used for cold water supply lines and for DWV. In most areas it is a bright white color with code markings on the sides. Sometimes codes specify that different wall thicknesses are required for DWV use, and the pipe used for these codes may be green or light blue. PVC should not be used for hot water supply lines. PVC is cemented or glued together and can be cut with common woodworking tools.

CPVC can be used for both cold and hot water supply lines. It is usually a tiny bit more expensive than PVC, but is worked just as PVC is. The same joint cement is used on both. As mentioned, CPVC and its fittings are usually cream- or tan-colored so as to be

Fig. 3-10 PVC pipe at top and CPVC pipe at bottom. Both have temperature and pressure limits printed on them.

Table 3-3 Plastic Pipe Factors

Type	Use	Diameters, Inches	Std. Lengths, Feet	Joining
PVC	Cold water supply	$1/2$, $3/4$, 1, $1\,1/4$, $1\,1/2$, 2, 3, 4	10, 20	Solvent cement and PVC fitting
CPVC	Cold or hot water supply	$3/8$, $1/2$, $3/4$, 1	10, 20	Solvent cement and CPVC fittings
PE	Cold water supply	Typically $3/8$, $1/2$, $3/4$, 1, $1\,1/2$, 2	25, 50, and 100 coils	Barbed fittings and clamps
PB	Cold water supply	$3/8$, $1/2$, $3/4$, 1	25, 50, and 100 coils	Barbed fittings and clamps
ABS	Drains, waste, and vents	$1\,1/4$, $1\,1/2$, 2, 3, 4	10, 20	Solvent cement and ABS fitting

distinguished from PVC. You should never use PVC fittings with CPVC. CPVC pipe will not deform with hot water use, but you can get leaks and joint failure if you use PVC fittings on hot water lines.

When in doubt, read the label. Both PVC and CPVC should be clearly marked, as in Fig. 3-10, with pressure and heat limits.

ABS is used primarily for DWV in home applications. It is rigid, usually black, pipe. The pipe is usually clearly marked with code and pressure information and is worked with ABS fittings. It should not be used for supply lines inside homes.

PE and PB are both flexible tubing and are easily worked. One often finds molded PB plastics (usually gray) used as cold water supply lines for sinks and toilets. It is easier to work and shape than copper. PB is also sometimes found as hot water supply lines from the cutoff valves to faucets.

PE, which is usually black, is typically used for cold water supply lines. Common uses include as supply lines to homes, supply and intake lines in wells, lawn irrigation pipe, and lines to ice makers. PE should not be used for hot water applications.

Faucet and toilet hoses, as in Fig. 3-11, are neither true pipe nor tubing. Because they are laminated layers of plastic and either fiberglass or metal fibers, they are very strong and flexible. The screw connectors on the ends also have soft gasket material in them to make leakproof sealing quick and easy. Because the thread

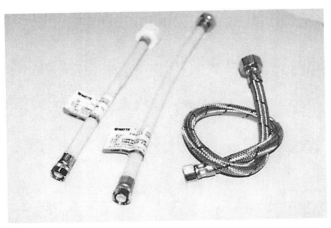

Fig. 3-11 *Plastic toilet and faucet hoses. Note the hose on the right is armored with steel mesh to resist high pressure.*

sizes of sinks and toilet connections differ, you can't use a sink connector on a toilet, or vice versa. They are available in lengths up to 36 inches and are ideal for both amateurs and professionals to use.

While they are perhaps a bit less attractive than chrome-plated copper or brass, they are less expensive and a lot easier to bend and to attach.

Clay and Fiber Pipe

Both clay and fiber pipe are used for sewer lines. Both are considered rigid pipe and are worked using fittings and sealants. *Vitrified clay* is a ceramic product baked to a hard, durable substance much like tile. It is hard, heavy, brittle, and difficult to work. Because of its weight, it is made in short lengths of 2, 4, and 6 feet for ease of handling. The short lengths make it more difficult to maintain a given slope in a trench, but it is impervious to most waste products and to corrosion from water or minerals.

Fiber pipe is usually made from glass fibers bonded with a bituminous resin. The advantages are that it is lighter in weight and the pieces are longer, which make it easier to slope in a trench. The pieces are usually 8 or 10 feet long. It is also much easier to work than clay. In a good many areas, both clay and fiber pipe have been supplanted by PVC. However, you still find them when repairing sewer lines in older homes.

In most places, repairs can be made with PVC and flexible sleeve coupling, as in Fig. 3-12. Terms used for this type of joint include sleeves, no-hub fittings, and banded couplings. These joints are made of thick, rubber or neoprene-coated fiber, much like the side of the tire of an automobile. You buy them in sizes that match the diameter of the pipe. The coupling is slipped down its full length on one pipe and then laid in place. Then when both pipes are in place, you slip the coupling up onto the second piece and tighten the clamps. This type of joint makes a lot of repairs quick and easy, but you should not use it on any pressured supply line.

FITTINGS

Fittings is a term that generically describes all the various pieces that are used to work the pipe into the desired direction and location. There are fittings for all types

Table 3-4 *Clay and Fiber Pipe Factors*

Type	Use	Diameter, Inches	Std. Length, Feet	Joining
Vitreous clay	Sewers and drains	4	2	Fittings and masonry cement
Bituminous fiber	Sewers and drains	4	8, 10	Tapered and tarred fittings

Note: Neither clay nor fiber pipe should be used within 5 feet of a house.

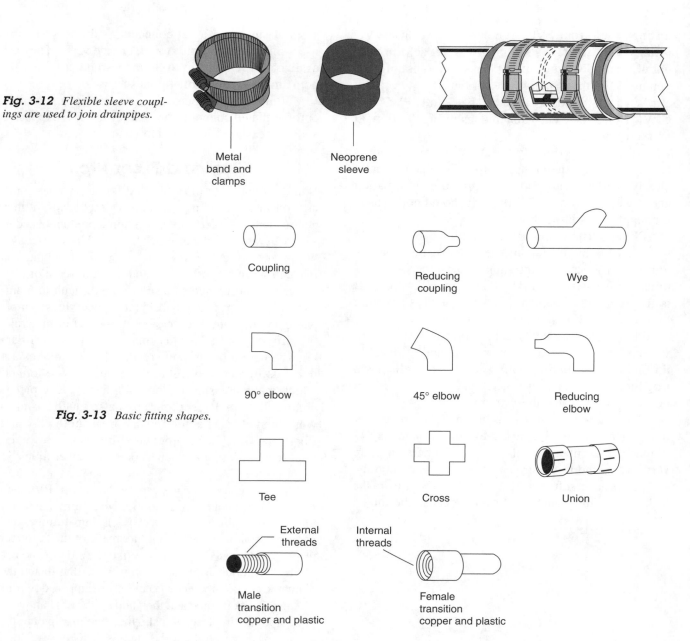

Fig. 3-12 *Flexible sleeve couplings are used to join drainpipes.*

Metal band and clamps

Neoprene sleeve

Coupling

Reducing coupling

Wye

90° elbow

45° elbow

Reducing elbow

Fig. 3-13 *Basic fitting shapes.*

Tee

Cross

Union

External threads

Internal threads

Male transition copper and plastic

Female transition copper and plastic

of pipe, and it is almost mandatory that the fitting used match the type of pipe. In other words, if you use galvanized pipe, you should use galvanized fittings. If you don't, you are vulnerable to all kinds of problems from corrosion to not passing inspection.

There is a wide range of fittings, as in Fig. 3-13. Similar shapes are used for steel, plastic, and copper pipes. The range of fittings for cast iron, clay, and fiber pipes is more limited, but 22½°, 45°, and 90° elbows are used as well as tees and wyes. One other point should be made. Fittings for DWV are different from regular fittings. Because they must carry a large volume of waste and water at a low pressure, they are designed to provide a more gradual and even change in direction than are regular fittings. See Fig. 3-14. DWV fittings

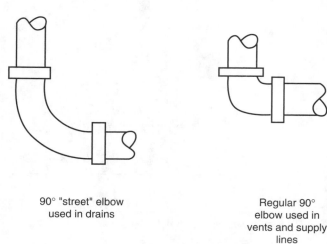

90° "street" elbow used in drains

Regular 90° elbow used in vents and supply lines

Fig. 3-14 *Drain fittings have longer and more rounded curves.*

are available for every kind of DWV pipe, such as ABS, PVC, cast iron, clay, and fiber. However, the type of fitting should match the type of pipe. ABS fittings should be used only on ABS, cast iron for cast iron, and so forth. The reasons are that the cements and sealants for each type are not interchangeable. You can't join ABS pipe with PVC cement.

The basic group of fittings for galvanized pipe is shown in Fig. 3-15. You can work pipe into almost any location by using just these fittings. However, there is a drawback to using galvanized fittings. A sealant must be used on the threads such as a pipe dope or Teflon tape. The pipe must be assembled sequentially and laboriously screwed in place. Heavy pipe wrenches must be used, and the bulkiness of the fittings makes their use in tight spaces difficult. Steel fittings also cost more than other types because they must be threaded at the factory.

The basic group of fittings for plastic pipe is shown in Fig. 3-16 while the basic group of fittings for plastic tubing is shown in Fig. 3-17. Figure 3-18 shows how clamps are used to hold the barbed fittings in place in both PE and PB plastic tubing. Rigid copper pipe fittings are shown in Fig. 3-19. Both copper and plastic fittings are smaller than steel fittings, making

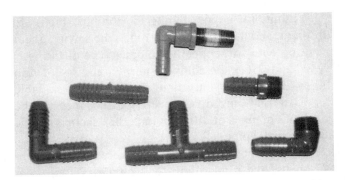

Fig. 3-17 *Basic barbed fittings used on PE and PB plastic tubing. Shown at top are female threaded adapter, middle coupling, and male threaded adapter; at bottom, 90° elbow, tee, and 90° male threaded adapter.*

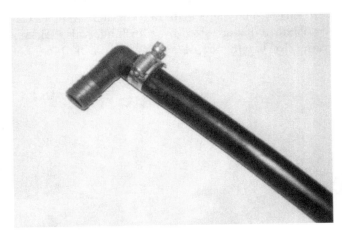

Fig. 3-18 *A 90° elbow clamped in PE tubing.*

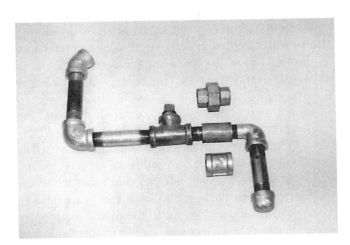

Fig. 3-15 *Basic fittings for galvanized pipe. From left to right, 45° elbow (ell), 90° elbow, tee, plug, union (top), steel coupling (middle), malleable coupling (bottom), 90° street elbow, and cap.*

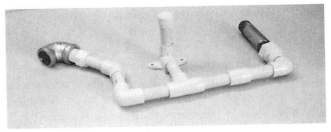

Fig. 3-16 *Basic fittings for plastic pipe are, from left to right, male thread adapter, 45° elbow, 90° elbow, tee, winged drop elbow, and female threaded adapter.*

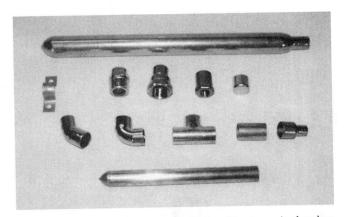

Fig. 3-19 *Basic rigid copper pipe fittings. Top is an air chamber. Second row, from left to right: strap hanger, male threaded adapter, female threaded adapter, threaded reducer, and cap. Third row: 45° elbow, 90° elbow, tee, coupling and reducer coupling. Bottom is a stub. A stub can also be used as an air chamber.*

them easier to use in tight spaces. Plastic fittings are the easiest to use as they are quickly cemented or welded in place. Copper fittings must be soldered, which involves both heat and time. Soldering fittings also requires some skill.

Fittings for copper tubing (soft copper pipe) are different, and they will be covered in the section about working with soft copper. One other type of fitting is very handy. It is the compression union, as shown in Fig. 3-20. Compression unions are commonly available for plastic, galvanized, or copper pipes, and they may be used to join pipes of different materials such as plastic to galvanized pipe.

You can move pipe almost anywhere from any point if you use the right fittings and nipples. Nipples are simply short pieces of pipe, as seen in Fig. 3-21. You can buy threaded nipples for steel pipe in lengths ranging from 1 inch to 18 inches from most vendors. They are normally sized by the inch up to 6 inches and by 2-inch intervals beyond that length. In some areas, nipples can be obtained that are sized by ½-inch intervals. A lesser variety of lengths are obtainable for plastic pipe, but the sizes are limited for copper pipe. The reason is that short pieces of copper or plastic pipe are easily cut and cemented or soldered. No unusual tools are required.

However, to cut your own short pieces of steel pipe, you must have the dies used to thread the pipe. Steel pipe is much harder to cut than either copper or plastic pipe, and pipe dies are very expensive. Threading pipe with manual dies is not an easy task. Muscles are required. It is much easier for even the professional plumber to combine nipple lengths than to stop, cut, and thread. You can get powered pipe dies, but the cost goes far beyond what is reasonable for anything less than professional use. Many licensed plumbers don't even carry pipe dies with them anymore.

VALVES

There are three types of valves commonly used as main supply valves in home plumbing: (1) globe valves, (2) gate valves, and (3) ball valves. Each type has different construction details and advantages. Figure 3-22 shows a cross section of each type. There are several other types of valves, but these are probably the most common.

There is one other type of valve that is usually incorporated into some other system. It is the *check valve,* as shown in Fig. 3-23. Most feature a flap that allows water to flow in one direction. When the water attempts to flow in the wrong direction, the flap seats against the opening. This will shut off the water flow.

Globe valves are the most common type and the least expensive. These use a washer to shut off the flow of water, as shown in Fig. 3-22A. You find these used as cutoff valves for kitchen and bathroom faucets, toilets, and hose bibs (that's the official plumbing name for the faucet outside the house). Globe valves are usually made of brass and are often chrome-plated for appearance. Because they use a soft washer for the seal, they may require maintenance from time to time.

Gate valves feature a sliding wedge that cuts across the full diameter of the pipe. See Fig. 3-22B. These valves are usually made of bronze. The wedge and its slot are usually machined and mated so that the pressure of the water will tighten and seal the valve in the cutoff position. These valves are larger and longer than check valves and are also more expensive.

Ball valves, as in Fig. 3-22C, also use the pressure of the water to effect a seal. However, the ball valve (usually a ball with a hole through it in one direction) is a simpler mechanism and saves space. A ball valve is smaller than either gate or check valves. Ball valves may be made of bronze, brass, PVC, stainless steel, or a

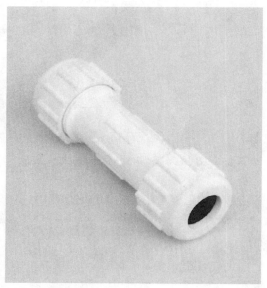

Fig. 3-20 *A plastic compression union, used to join plastic pipes, or plastic to metal.*

Fig. 3-21 *Nipples are short pieces of pipe used to put fittings together.*

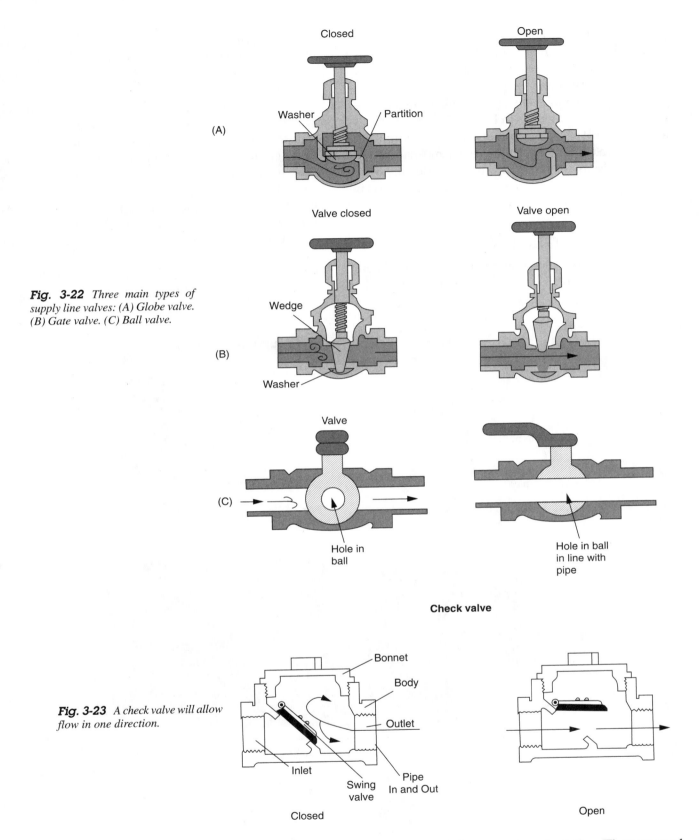

Fig. 3-22 *Three main types of supply line valves: (A) Globe valve. (B) Gate valve. (C) Ball valve.*

Fig. 3-23 *A check valve will allow flow in one direction.*

combination of these materials. Ball valves are relatively inexpensive.

Saddle valves, as seen in Fig. 3-24, are used to connect low-flow auxiliary components to existing water lines without doing extensive adaptation. They are used primarily on rigid copper pipes. The valve is clamped firmly across the pipe, and the needle valve is screwed in to puncture the pipe. The plastic seals around the junction of the valve and the pipe provide a seal to prevent leakage. The most common application is to connect a flexible copper or plastic line to the ice maker of a refrigerator.

There is one more valve that should be mentioned. It is the *sill cock,* as shown in Fig. 3-25. The sill cock is used in place of a hose bib in cold climates where outside access to water is needed, but freezing temperatures will not allow the use of standard valves or hose bibs. Variations of the sill cock are used as all-weather faucets in yards and farm grounds.

GROUNDS

There is one more plumbing-related item you should consider. Metallic pipes are often connected to the electrical distribution panel as an electrical safety precaution. A ground clamp, as shown in Fig. 3-26, is used. Metal pipes that are buried are grounded because they are buried.

To ground a pipe that isn't buried, a rod or a heavy wire is driven deep into the ground where the soil stays moist. Then a ground clamp is used to connect the pipe to the ground wire. This clamp should have a dielectric plate to prevent electrolytic action between two different metals. The clamp can be used to ground a galvanized pipe, for example, to a copper ground wire. This is only needed in one location for a house.

There is one more grounding application. If, for example, you repair a leak in a copper pipe with a piece of plastic pipe, you may have disrupted the continuity of the grounded pipe. To maintain the integrity

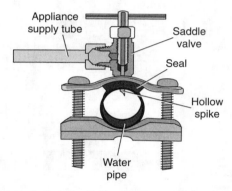

Fig. 3-24 *A saddle valve is used to connect appliances such as an icemaker.*

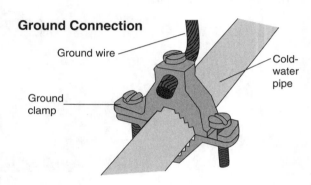

Fig. 3-26 *A ground clamp is used to connect the electrical ground.*

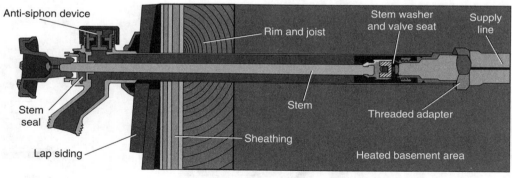

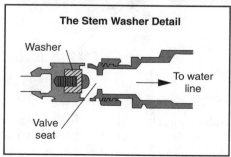

Fig. 3-25 *A sill cock is a special faucet. The valve is in the heated basement area and doesn't freeze in cold weather.*

of this ground, use two clamps and a wire to connect the two metal sections that are separated by the plastic patch, as in Fig. 3-27.

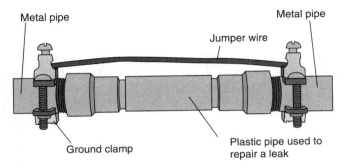

Metal pipe

Metal pipe

Jumper wire

Ground clamp

Plastic pipe used to repair a leak

Fig. 3-27 *Continuing a ground. This is an important safety factor.*

MAKE ESTIMATES

Once you know a bit about codes and regulations and exactly (or close, anyway) what you want as an end result (add a new bathroom, replace your main supply line with a larger line, get rid of rusting pipe, etc.), you can begin to develop the estimates. Most of us think of estimates as a cost projection. However, there are several other factors to consider. These include how long the job will take, who does the work, and at what stage should inspections occur.

Extent of work The extent of the work simply means how much and what type of work will it take to finish the job. Will you have to tear into a wall? Dig a trench? Connect one pipe or several? Can you access the spot easily, or is it hard to reach and cramps your movement?

This is one of the situations where a lot of amateur plumbers get in a bind. For example, one fellow thought he could add a shower to a tub unit by simply changing the faucet housing behind an access panel. That's not a real easy job in itself. But the guy didn't think and tried to get a long pipe through a short opening up to a hole he had drilled in the wall. This was after a considerable expansion of his vocabulary while changing the faucet housing. He forgot about screwing the fittings on in sequence and a lot of other things. First, he tried using short nipples rather than a single, long pipe. It almost worked. Eventually, he wound up tearing out part of the wall, installing the pipe, repairing the wall, installing a water-resistant panel, and finally adding the external fixtures and shower head. He also lost a lot of time with repeated trips to a hardware store. It was a long weekend for him. A little realistic planning and an estimate of the scope of the work would have saved him a lot of time and energy.

A checklist, such as that in Fig. 3-28, is helpful in determining the scope of work. To make one, you simply list what you think the steps are. Then you can make columns in which you can list the materials needed and the time to do the job. You can also list where to stop for inspections, if that's needed.

Material requirements If you take a little time to list what you will need, you can save a lot of time and frustration with repeated trips to a vendor. A good list will help you remember in the store as well as give the salesperson a good idea of what you are doing. They can sometimes offer good ideas and can also tell you where to find the stuff you're hunting. Questions to ask yourself include these: Do I need to change directions? How many times? What fittings do I need, and how much pipe? Do I have the right tools? Do I have enough gas for my torch? Double-check the sizes you will need, too.

If you tear into a wall, you will need materials to repair the wall as well, so you might as well get that too. From experience, last year's plastic pipe cement or drywall spackle is probably too dried out to use this year. Consider everything you will need to do the job, and check to see if you have enough supplies and tools present (and not on loan to someone) and in good working order.

A final idea to consider is that a lot of experienced amateurs always buy a few extra fittings when they purchase the materials. The cost is usually rather small. However, if you ruin one, an extra one on hand can save you a lot of lost time and frustration in getting a new one (especially late at night). A lot of us stockpile these for later jobs, and often we find that we can finish a job without a special trip.

Costs Your costs are the sum of all the parts. However, there are supplies such as solder, cement, torch fuel, pipe sealants, and so forth that are sometimes overlooked. If you haven't checked on prices for a while, you might do that too, because they haven't gone down lately. If a major fixture or appliance is involved, such as a hot water heater, comparison-shopping is very wise. Prices for the same or a similar item can vary widely, sometimes by hundreds of dollars. You might also consider that the cost of professionals and inspection fees may be involved.

Time requirements This is another point where a lot of people make mistakes. What looks like a quick and simple job usually takes longer than you think. How long does it take to unsolder a large fitting? If you are going to replace it, how long will it take you to uncrate the fitting and prepare it to be worked? Once you have outlined the

CHECKLIST
SCOPE OF WORK

Problem

Cause of problem

Parts involved

Materials needed

Tools needed

    On hand

    Purchase

    Rental

Safety gear needed

Steps   **Time needed**

Inspections

    Required

    When

    Delays/scheduling

Fig. 3-28 *A scope of work checklist.*

steps in your list, you can make some estimate of how long each will take. Simply total your estimates, and you can get a more accurate idea of how long the job will take.

Who does the work? Many places allow the home-owner to do extensive work. Others, however, give them little latitude. If you have checked and know what you can do and what you can't, then you can pinpoint the places where you must stop for professional work or for inspections.

Inspections required When planning inspections, you should remember that inspectors' schedules are generally pretty full. It may take several days for the inspector to show up. You should try to take into account some delays and whether someone needs to be present to give the inspector access. Rough-in plumbing usually requires a pressure test where the inspector plugs the system, pumps it full of air to a pressure of several pounds per square inch, and then waits to see if the pressure drops. The process requires at least half an hour. Inspectors will usually answer questions and provide a considerable amount of useful information.

Scheduling In making estimates for your project, you may not need to consider scheduling unless you have to arrange for several different people to do different jobs. If, however, you must arrange for a professional plumber, an inspector, a plasterer, and a painter to each do separate parts of the job, then you had best consider schedules.

This is one of the biggest jobs (and headaches) of building contractors. It is a real task to get all parties to arrive just when they should and to finish on time. You should expect delays in arrival and longer times to do the job. If you have to deal with this situation, it is best to schedule at least several hours, if not a couple of days, between each of these.

ADJUSTING YOUR WATER PRESSURE

Most communities have acceptable water pressures. However, in some places the water pressure is more than 80 pounds per square inch (psi). In others, the pressure may fall below 25 or 30 psi. Occasional peaks or slumps in pressure should be of little concern. But if your community has constant pressures that are either too high or too low, then you should consider adjusting your pressure.

Devices called *regulators* are used to reduce pressures, while special pump and air chamber units can be used to boost the pressure. In some communities, the building codes may require the installation of these units.

Regulators As mentioned, regulators are used to reduce pressures that are too high. They work on the venturi principle of using a small input or supply opening to allow the high-pressure stream to enter a larger chamber. When a stream passes through a small opening into a larger opening, the pressure is reduced. Also, if the water pressure rises, the increase will force the thin diaphragm upward, which will close the outflow valve even more and thus reduce the pressure to offset the increase. The outflow from the chamber is at a lower pressure. Most regulators have a diaphragm in the chamber that allows the exact pressure to be adjusted. See Fig. 3-29.

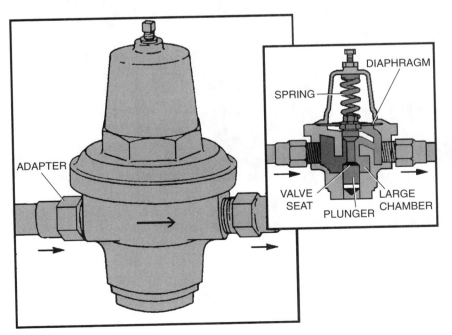

Fig. 3-29 *A regulator can lower water pressure.*

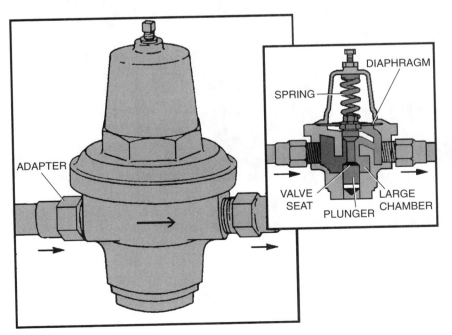ADAPTER

DIAPHRAGM

SPRING

VALVE SEAT

PLUNGER

LARGE CHAMBER

Booster pumps Booster pumps are used to boost low-pressure systems. Figure 3-30 shows how a pump forces the water into the tank and compresses the air at the top of the tank. The compressed air then boosts the water pressure. Such a system typically has a limit switch that turns the pump on when the tank pressure drops below a certain point. The limit switch will also turn the pump off when a certain pressure is reached.

PREVENTING WATER CONTAMINATION

Vacuum breakers Vacuum breakers are now required by some codes on all hose bibs and sill cocks. They are small devices that can be screwed between the hose and the hose bib.

When the water pressure in the main line drops suddenly, such as when a water main breaks, the main line will act as a siphon to pull water in the supply pipes of the various home systems connected to it. If a water hose had been in use to flush a clogged drain, or with a garden spray unit spreading insecticides, then the siphon action on the water main would pull the drain waste or the insecticide back into the potable water system.

The vacuum breaker is constructed so that it will let air into the system when a reverse pressure is detected. Don't confuse the vacuum breaker with a backflow preventer.

Backflow preventers Backflow, or backup, preventers are devices installed in drains. If a sewer main floods and the water level in the sewer rises above the level of the drains to a home, then the sewer will flow back into the house. This happens occasionally where a house with a basement is located in a low-lying area. It also happens when an area is flooded or is swamped with torrential rainfall.

If you are in a situation where main sewer lines periodically flood due to heavy rains or floods, then backflow preventers are a good idea. They simply function as check valves that allow flow from the home to the sewer, but block flow coming from the sewer.

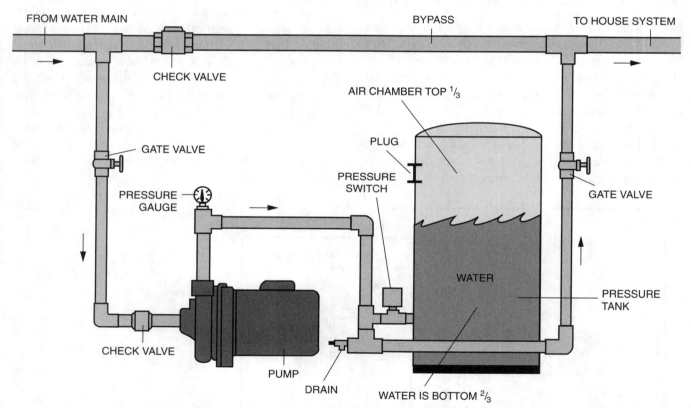

Fig. 3-30 *A booster pump system can maintain a higher water pressure.*

4
CHAPTER

Pipe Working and Plumbing Tools

THIS CHAPTER FOCUSES ON THE VARIOUS TOOLS that are used to work pipe. There are several categories of these tools. First, there are tools that are pretty common and are used in a variety of applications, including plumbing. This group is usually called *general-use* or *general-purpose* tools. Next are some tools that are used only in plumbing and repair work, but aren't used to work pipe. Finally, there are tools designed to shape and work each type of pipe, such as steel, plastic, or copper pipe.

GENERAL-USE TOOLS

General-use tools, or general-purpose tools, are those basic tools that just about everyone should have. They are common from jobs on automobiles to furniture repair and, of course, plumbing. They can be found in the toolbox of a professional as well as that of an amateur. With just these general-use tools, you can often do a lot of plumbing jobs, such as fix leaky faucets or tighten a loose pipe or fitting. Most of these tools are hand tools, but there are several common power tools that should be considered.

To help sort out these tools, we've divided them into three groups: (1) gripping and turning tools, (2) cutting and demolition tools, and (3) layout and measuring tools. There is also a section on common power tools as well as some ideas on how to rate them. If you don't have these tools and you do a lot of work around the home, it wouldn't hurt to pick these up.

Gripping and turning tools These are tools that are used to drive screws and tighten bolts or fittings. Figure 4-1 shows some of these common tools.

First, you should have several screwdrivers. You should have large and small sizes of regular slot head and Phillips head screwdrivers. The four-in-one screwdriver, in Fig. 4-2, is a popular and inexpensive tool because it has large and small sizes of both types of screwdriver. There are also other types of screwdrivers as well, such as the star and square head types. However, just a slot head and Phillips head will usually do.

Next, you will need some good pliers. The three main types used in plumbing are the regular pliers, needle-nose pliers, and the channel lock variety, as seen in Fig. 4-1. It doesn't hurt to have both large and small sizes in these, too, but one of each works pretty well.

Next, you should have several types of wrenches. Because plumbing fittings are usually larger than the wrenches found in most standard wrench sets, a large adjustable wrench is advisable. Two are better than one. You will also need two pipe wrenches, as shown in Fig. 4-1. It's okay to have a large one and a small one, but you will need at least two. The reason is that you use one wrench to hold pipe in place while you use the other to tighten things. More on this later.

Fig. 4-1 *Common gripping and turning tools. At top left is an Allen wrench set. At top is a set of combination wrenches, a four-in-one screwdriver with large and small slot and Phillips heads, a socket wrench set and ratchet handle, and regular, needle-nose, and channel lock pliers. Large and small combination and pipe wrenches are also needed.*

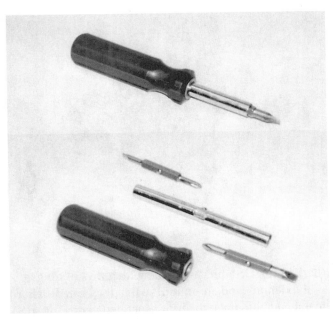

Fig. 4-2 The four-in-one screwdrivers are very versatile. At top it is ready to use. The bottom shows the large and small slot and Phillips heads.

There are also a variety of assembly and disassembly operations that fall within normal ranges, so a good set of either combination or open-end wrenches, such as in Fig. 4-1, is also advisable. Finally, a set of both standard and metric Allen wrenches, also shown in Fig. 4-1, is useful. A lot of handles and collars used in plumbing are fixed in place with Allen screws.

Cutting and demolition tools Typical cutting and demolition tools, such as in Fig. 4-3, include saws, chisels, hammers, and nail bars. Tool sets should include saws for sawing wood and plastics and a hack saw for cutting metal. The chisel collection should include at least one wide wood chisel and a narrow one. It should also include a cold chisel for cutting metal. You will need at least one hammer for banging. The most common hammer found is a claw hammer. This is the main carpentry hammer, and it is not recommended for banging on metal. To bang on metal, you should use a ball peen hammer, which has one end rounded. See Fig. 4-3. The face of the ball-peen hammer is usually made of tougher steel and is tempered differently from the face of a claw hammer.

A lot of people make a mistake when they buy a lightweight hammer. A lightweight hammer does not give you much impact power. You have to hit a nail, or anything else, about twice as much and about twice as hard to do the job. A 16-ounce claw hammer or ball-peen hammer is a good, functional weight.

Layout and measurement tools This group of tools, as seen in Fig. 4-4, is used to measure distances, set fixtures, and determine slopes and locations. A good tape measure or folding rule is a must. It should have numbers that contrast with the background so that they are easy to read. A plumb bob is useful to find true vertical and to locate a spot directly beneath something.

A level is used in leveling fixtures such as sinks and lavatories. It is also used to set things so that they aren't level, such as drainpipes and waste pipes. They have to slope in order to drain away the stuff inside the pipe. A chalk line is useful in marking straight lines over a distance. A square of some sort is also essential. You can use a combination square, a carpenter's square, or a

Fig. 4-3 Common cutting and demolition tools include a wood-cutting hand saw, a hacksaw, cold chisel, wide and narrow wood chisels, ball-peen hammer, and claw hammer. A nail or ripping bar is also recommended.

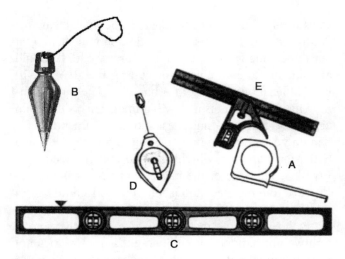

Fig. 4-4 *Common layout and measuring tools include (A) tape measure, (B) plumb bob, (C) level, (D) chalk line, and (E) combination square.*

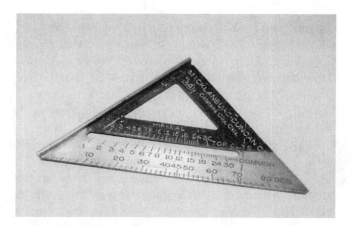

Fig. 4-5 *A carpenter's speed square has angles marked on it.*

carpenter's *speed square,* as seen in Fig. 4-5. The speed square has various angles marked on one side, which comes in handy when you need to lay pipe in an angle other than 90°.

Finally, you will also need some marking tools. These will include a pencil, some ordinary chalk or welder's chalk, and something to scribe metal. This could be a scratch awl or the tip of dividers. You might try a ballpoint pen, too. Sometimes a ballpoint pen will work as well as an awl. For plastic pipe a fine-point felt-tip pen works very well.

Common power tools There are several common power hand tools. The most versatile of these is the power hand drill, as seen in Fig. 4-6. You can use this to drill holes in wood or metal with the appropriate drill bits. A word about bits is in order. We recommend that you buy only drill bits that are marked *HSS* (for high-speed steel). It's also a good idea to get HSS bits that are coated with either titanium (gold color) or cobalt

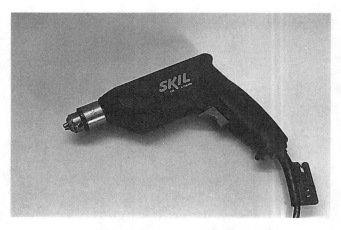

Fig. 4-6 *A power hand drill.*

(dark blue). HSS bits are made with alloys of tungsten, molybdenum, and chromium so that they can drill hard metal and hold up to the heat generated in drilling. HSS bits will drill wood or metal. The coatings such as titanium and cobalt just make them a little tougher. Bits that aren't marked or are marked *tool steel* probably won't drill metal.

You may also need special bits to drill concrete and to drill large holes in wood. These are shown in Fig. 4-7 and are concrete or masonry bits, hole saws, and spade bits. You can buy a hole saw, as shown in a variety of sizes. You can also buy hole saw sets that have several different-diameter blades that can be mounted on one drive spindle.

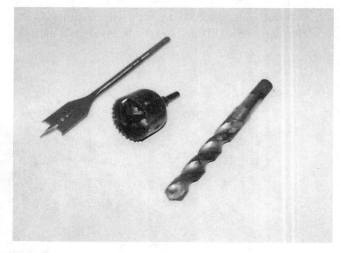

Fig. 4-7 *You may also need, from top, spade bits, hole saws, and masonry bits.*

Battery-operated power tools, such as the drill in Fig. 4-8, are both popular and handy. They don't have power cords to tangle you up, and they don't require a chuck key to tighten bits in the chuck. They are heavier and the

Fig. 4-8 A cordless hand drill with variable torque.

Fig. 4-10 A power screwdriver with a variety of bits and sockets is very versatile.

batteries must be periodically charged. The chuck has several torque settings that allow the bit to stop turning when a given resistance is attained. You can use a variety of bits that include slot, Phillips, star, Allen, and square driving bits as well as sockets and other turning tools.

There are a variety of battery-operated tools available. Figure 4-9 shows a circular saw and a drill that operate on the same type of battery. This allows both batteries to be used on either tool, or both to be recharged with a single battery charger. When you buy battery-powered tools, note that the higher voltage ratings usually provide more power than lower voltage ratings. Professional-grade tools are about 3 times more expensive.

Fig. 4-9 Cordless power tools are very convenient. This group includes a drill, a rotary saw, and a reciprocal saw, all very useful for plumbing.

Another handy tool is the power screwdriver, shown in Fig. 4-10. This tool provides more rapid action than a hand tool with a lot less exertion. It also has a chuck that allows for rapid change of bits, so that you can drive slot, Phillips, square, or even Allen head screws with ease and speed.

Other power tools that are useful in doing plumbing work are a saber saw (often called a jigsaw) and a reciprocal saw, as in Fig. 4-11. Both of these tools can now be obtained in battery-powered, or cordless, versions.

SAFETY

Working safely is the single most important skill. It's better to go without water for a while than to be without an eye for a lifetime. Scary? Yes, but it can happen if you aren't careful.

First, you should first develop both an attitude of protecting yourself from the hazards of the job and the habit of using protective gear. If you are soldering copper pipe, the hazards will include touching hot metal and drops of molten solder falling into an eye or on unprotected skin. The smart thing to do is to wear gloves, goggles or a face shield, and long sleeves. If you are using a flux that causes fumes, then a face mask is also a good idea.

If you are using a torch or a small furnace to melt lead or solder, it is a very good idea to have a home fire extinguisher handy. You probably won't need it, but if you do, it could be a home saver or even a lifesaver.

Every job, even the smallest one, has a risk. There are hazards from sharp edges, heat (hot metal usually doesn't look hot), fumes, and unclean material. The items shown in Fig. 4-12 should be part of your tool kit. Heavy work gloves that extend past long sleeves, face masks that filter the air you breathe, and eye protection are recommended in the strongest terms. Another idea to consider is heavy work clothes, or a work apron. The work clothes should be made from natural fibers and have long sleeves. The reason is that most synthetic fibers will either rapidly burn or melt. Sturdy boots or shoes will help as well because you may be walking

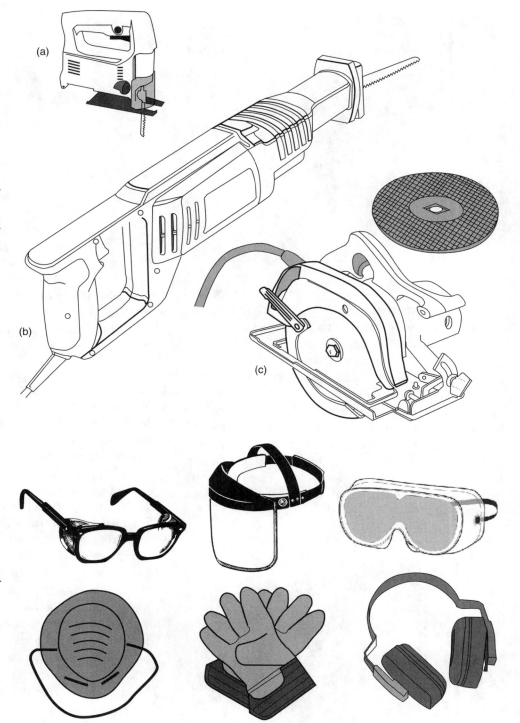

Fig. 4-11 *Power saws such as the (A) saber saw, (B) reciprocal saw, and (C) circular saw are useful. Wood and metal cutting blades are available for each.*

Fig. 4-12 *Safety items are a must. Include eye protection, gloves, filter masks, and noise suppressors.*

over sharp or hot bits of metal and sometimes in wet areas. Finally, some jobs are very noisy. Earmuff-type noise protection or soft earplugs are also good ideas. Remember, if you don't protect yourself, who will?

SPECIALTY TOOLS

There are several different categories of specialty tools. Some are used on different pipe types to unclog pipes. Others are used on faucets, sinks, drains, and filters. There are even some laser tools that are useful in finding level lines or sloped ones. There are still other tools used in special cases to cut pipe. To top all this off, there are also special tools used on certain pipe types. These specialty tools are grouped as the general-use tools are, except that the special tools for a given pipe type will be shown in a pipe group.

Tools to remove clogs   This group of tools is probably the most common and most often used group. The ubiquitous "plumber's friend" (the plunger) is every homeowner's first line of defense. There are three basic types. The first is a sink plunger, and it is flat across the bottom. The flat bottom will seal on the flat bottom of a sink. The second type is a toilet plunger, and it has a cone on the bottom. The cone is better suited to seal against the bottom of a toilet. The combination plunger shown in Fig. 4-13 has a cone that will fold up inside the bell to leave a flat surface. It can be used for both sinks and toilets. This type is recommended for the typical homeowner. The plain bell on a plunger works very well, but some with expanded bellows can give extra force to the plunger action.

Then there are the *snakes*. These are the drain augers. The one shown in Fig. 4-13 is recommended

for homeowners. Its diameter of $1/4$ inch gives it great flexibility, and a length of 25 feet provides good *reach.* Larger diameters and lengths are needed to clean out sewer lines. Some can also be used with power tools.

Two other tools shown can be added to your toolkit for little expense. The first is the *closet auger,* shown in Fig. 4-13. Remember that toilets are officially called *water closets.* Now, some of these tools are marketed as *toilet augers* as well. The ceramic material in a toilet is easily marked by iron or steel. The closet auger provides a tube and a plastic sleeve at one end to allow the auger to be inserted into the trap area of the toilet without the auger touching the ceramic.

The last tool in Fig. 4-13 is a blow bag. It is attached to a water hose and inserted into one end of the clogged pipe. When the water is turned on, the flexible walls of the blow bag "blow up," or expand, to seal off the pipe

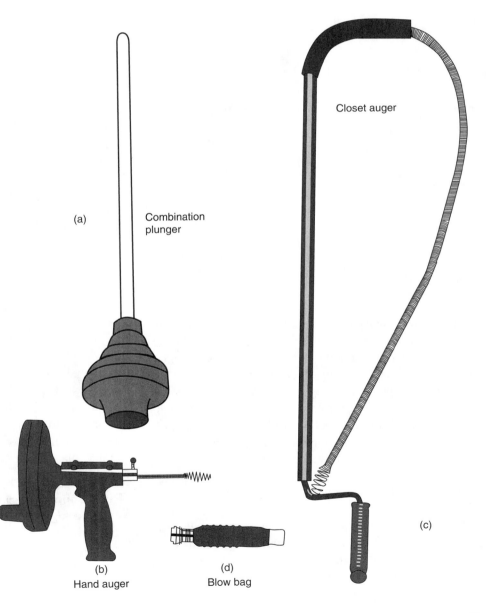

Fig. 4-13 *Tools to remove clogs. (A) Combination plunger. (B) Drain auger. (C) Closet or toilet auger. (D) Blow bag.*

(a) Combination plunger

Closet auger

(b) Hand auger

(d) Blow bag

(c)

at the top. The force of the water in the hose is directed fully against the clog.

Gripping and turning tools   This category contains several tools, as shown in Fig. 4-14. A strap wrench is used to grip large, round shapes, such as drainpipe and waste pipe and filters. A spud wrench is used on traps, and a faucet handle puller is handy for prying off stubborn, mineral-encrusted faucet handles. Seats in faucets can be unscrewed with a seat wrench, and a basin wrench can be used to reach up behind a lavatory sink to unscrew the nuts holding the faucets in place. The last wrench in Fig. 4-14 is used on filters.

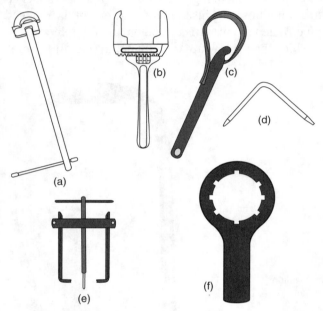

Fig. 4-14 *Special gripping and turning tools. (A) Basin wrench. (B) Spud wrench. (C) Strap wrench. (D) Seat wrench. (E) Handle puller. (F) Filter wrench.*

Cutting and demolition tools   This tool group in Fig. 4-15 includes pipe cutters, tubing cutters, plastic tubing cutters, chain cutters, small hacksaws, burring reamers, and metal cutting blades for power tools. The cutters for pipe and tubing are essentially the same type of mechanism. A rotary cutting blade is forced just a little bit into the metal. The tool is rotated, and the cutter is forced a bit more into the metal and rotated again. The process is repeated until the pipe is cut. Pipe cutters are 12 to 18 inches long while tubing cutters are only about 4 or 5 inches long.

A chain cutter is sometimes used on large-diameter cast iron pipe. There are several cutters fastened into the chain that will cut when the chain is tightened. The flexible metal chain allows the cutter to be used on several large diameters. However, this expensive tool

is rapidly being replaced by metal cutting blades used in power saws. More will be said about that later.

After a pipe is cut with a rotary pipe cutter, there is a sharp edge called a *burr* inside the pipe. This burr can reduce the effective diameter of the pipe, so it is removed. A pocketknife can be used on plastic pipe, but a reamer must be used on metallic pipe.

Plastic tubing is often cut with a special tool. It is very much like a pruning shear, except that it has a compound lever action that allows large diameters to be cut with minimal effort. It is a moderately expensive tool, and the amateur can do without it if the scope of the job is limited.

Finally, a small hacksaw, as shown in Fig. 4-15, is a very handy thing to have. You can reach into tight spaces and even between the sides of a narrow opening. It will allow you to cut in areas you normally couldn't reach with a regular saw.

Layout and measurement tools   Most pipe working jobs won't need anything more than standard layout tools. However, one innovation has made a number of jobs quicker and easier. Laser levels, while not very new, are now quite inexpensive. The average amateur can easily afford to own one, and they are indeed handy. A laser level, such as that in Fig. 4-16, can project either a point or a line. You can use these to level an object, to project a spot or line level with something, or to project a slope.

The last special tool is a stout piece of string and a string level. A string level clips onto a taut string, as in Fig. 4-17. This lets you use a simple string to project levels or slopes. If you need to swing an arc, or draw a large circle, you can use the string as a compass, as in Fig. 4-18.

SPECIAL TOOLS FOR EACH PIPE TYPE

There are a few special tools used to work galvanized steel pipe and both types of copper pipe. Plastic pipe is easily worked with wood- or metal-cutting tools. It doesn't require heat or wrenches to set it in place, so no special tools are required. If you decide to do a job with galvanized steel or with either type of copper, you should plan ahead and buy, borrow, or rent what you need.

Rigid copper special tools   Rigid copper pipes are soldered together. To solder pipe, you need the tools shown in Fig. 4-19. You will need a propane torch, a spark igniter, solder, flux, an acid swab for the flux, and steel wool or emery cloth. It's also a good idea to have a wire brush, a damp rag, and a piece of bread.

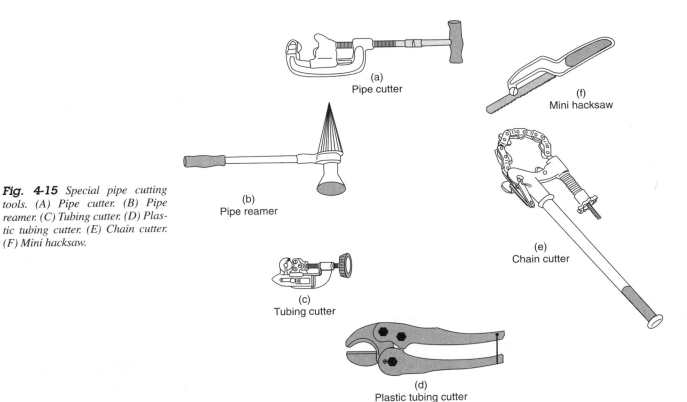

Fig. 4-15 *Special pipe cutting tools. (A) Pipe cutter. (B) Pipe reamer. (C) Tubing cutter. (D) Plastic tubing cutter. (E) Chain cutter. (F) Mini hacksaw.*

(a)
Pipe cutter

(f)
Mini hacksaw

(b)
Pipe reamer

(e)
Chain cutter

(c)
Tubing cutter

(d)
Plastic tubing cutter

Fig. 4-16 *A small laser level can project a horizontal line, a vertical line, or a dot. By adjusting the laser level on the tripod, either a level or a slope can be projected.*

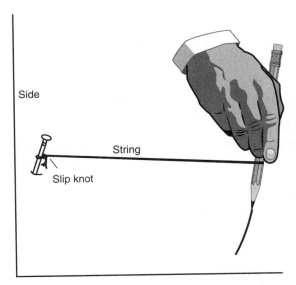

Fig. 4-18 *A string compass makes quick and easy circles or arcs of any size.*

Side

String

Slip knot

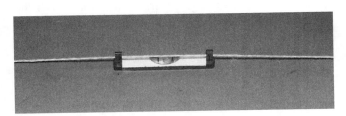

Fig. 4-17 *A string and a string level can be used to define a level or sloped line.*

To make a soldered joint, both pieces must be very clean. You can use the wire brush and abrasive for cleaning. The flux helps keeps the heated area clean and free of oxides (copper forms an oxide very rapidly when heated) and helps the solder flow. The damp rag can be used to wipe off excess solder. The bread is used to stop up a pipe so that no water can reach the area being soldered. Any water present where you are

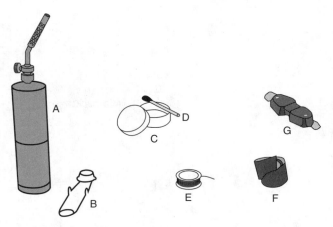

Fig. 4-19 *Tools needed to solder copper pipe are (A) propane torch, (B) igniter, (C) flux, (D) acid swab, (E) wire solder, (F) emery or sandpaper strip, and (G) goggles or face shield.*

trying to solder will cause the area to rapidly cool and probably ruin your effort. The water will dissolve the bread once you turn the water back on.

There are some things you should know about solder, too. Solder is an alloy of lead and tin. Solders are rated by a ratio with the percentage of lead being the first number. For example, 60/40 solder is 60% lead and 40% tin. Because tin is more expensive than lead, 60/40 solder is cheaper than 40/60 solder. However, 60/40 solder may be more difficult to work. Probably the best solder to use for copper pipe is 50/50 or 40/60 solder. The 50/50 solder is by far the more common of the two.

The most common shape of solder is wire solder. It also comes in bars and as paste. The paste is easy to use, but the bar solder is more difficult than wire solder. Paste solder is tiny bits of solder suspended in a thick solution of flux. You simply brush it on, assemble the joint, and then heat it until the solder melts and fuses. You can see this happen around the edges.

Wire solder is the most common shape for the amateur. It is available in solid, acid-core, and rosin-core forms. The core-type solders have a flux embedded in the center of the wire. For copper pipe, acid-core solder is recommended. Rosin-core solder is used with electrical wiring, and solid solder does not have a flux built into it and is sometimes a bit more difficult to work.

Lead in the solder is a possible contaminant for water. Over a long period of time, the effects of the lead could cause problems for a person. To offset this potential hazard, some codes now require the use of lead-free solder. It is an alloy of tin and several other metals and works just as regular solder does. It just doesn't contain lead. It is a bit more expensive.

You may hear the terms *hard solder* and *soft solder*. For joining copper or brass pipe, you use a soft solder like those just mentioned. *Hard* solder refers to using silver, brass, bronze, or copper as the solder. You may have heard someone talk about silver solder. It's more difficult, takes more heat, and is more expensive.

Flexible copper special tools Flexible copper tubing is easily worked and requires fewer fittings than rigid copper pipe. There are three special tools, however, that can make the job quicker and easier: tubing benders, swages, and flaring tools.

If you try to bend tubing by hand, it usually kinks on you. A kink restricts water flow and reduces the quality of your water supply. To prevent kinks, you should use a series of wire bending sleeves, as in Fig. 4-20. You use a different size for each size of tubing. The sleeve must fit snugly on the tubing. If you are bending sizes above $\frac{1}{2}$ inch, you can use a conduit pipe bender, also shown in Fig. 4-20. You can find these in the electrical conduit section of a building supply center.

Fig. 4-20 *Bending tools for copper tubing include bending coils, a conduit and tubing bender, and a bending jig for small diameters.*

Flexible copper tubing and rigid copper pipe have different outside diameters for a given size. For any given size (say, $\frac{1}{2}$-inch), pipe is larger than tubing. This means that you can't join copper tubing directly to copper pipe. You can buy special fittings that allow you to fasten one to the other. They are expensive fittings. If you have several joints to make, it is probably worth your money to buy a set of swages. These are hardened steel tools that are driven into the end of the tubing to enlarge it. See Fig. 4-21. Swages can be purchased in sets and in individual sizes, and they can be rented from some equipment rental stores.

Flaring tools are used to expand the ends of the tubing to match the shape of the fittings used on tubing. See Fig. 4-22. A watertight seal is formed by matching the shape of the flare to the shape of the fitting. The nut on the fitting holds the two pieces together.

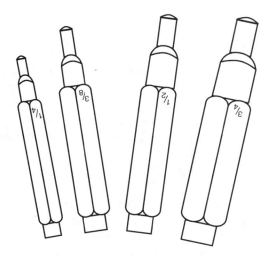

Fig. 4-21 *Swages are used to expand the end of copper tubing to act as a coupling for copper pipe.*

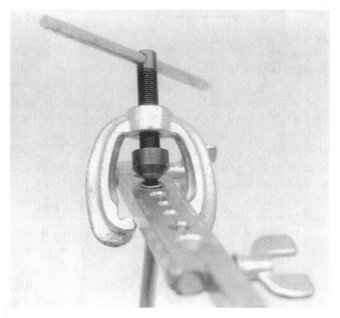

Fig. 4-22 *Flaring tool for copper tubing.*

Galvanized pipe special tools About the only special tool needed to do extensive or complicated work with galvanized pipe is for threading. Galvanized pipe is assembled by screwing together combinations of long pipe pieces, fittings that change the direction or size of the pipe, and short pieces of pipe. Each end of the pipe and each end of the fitting must have threads. If you cut the end off of a piece of galvanized pipe, you have cut the factory thread off it, and you either can't use it or must thread it.

Lengths shorter than 4 feet are usually called *nipples*. You can purchase nipple combinations to make up any number of lengths. However, if you are doing extensive work with galvanized pipe, you should consider renting or buying the pipe dies used to thread the ends of the pipe. This will allow you to cut pipes to exact lengths and thread them rather than use a variety of expensive fittings. Using a large number of fittings also takes a lot of assembly time and poses a greater number of leaky joints.

Figure 4-23 shows you what the pipe dies look like. The die holder or wrench features a housing and a handle. Each pipe size has a separate die. The die used to cut threads on ³/₄-inch galvanized pipe can't be used on 1-inch pipe. You exert a lot of torque on the pipe when you cut threads on it. So, you also need something to hold the pipe still while you thread it. You shouldn't use an ordinary vise because the jaws will distort the roundness of the pipe. Some vises have pipe jaws underneath the regular jaws, but the best tool is a pipe vise, also shown in Fig. 4-23. The pipe vise can be bolted to a workbench or to a stout board, such as a 2 by 8 about 4 feet long. The board can then be moved about and clamped to something to hold it steady.

A standard piece of pipe is 21 feet long and is called a *joint*. You can often purchase several joints of pipe from a steel supplier much cheaper than you can purchase threaded pieces of varying lengths from a building supply store.

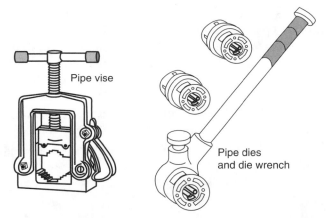

Fig. 4-23 *Pipe can be held in a pipe vise and threaded with pipe dies.*

Cast iron pipe tools There are several special things you will need to work cast iron pipe. The joints on cast iron pipe are sealed with melted lead, so you need tools to do that. These include a torch or furnace, oakum, a ladle, lead, tamping tools, and a joint runner. More will be said about these later. Cast iron pipe is difficult to cut, particularly if you are cutting away a damaged section of it in a basement. The best tool to use then is a chain cutter. Otherwise, you can cut it with hacksaws, circular saws with metal cutting blades, or reciprocal saws with metal cutting blades.

RENT OR PURCHASE?

If you don't have the tools you need to do a job, you have four choices. You can buy the tool, borrow it, or rent it. The fourth choice is to call a plumber. Unless you have a really well-stocked, super do-it-yourselfer neighbor, you probably can't borrow it either. That leaves most of us with two choices: buy or rent.

If it is rather inexpensive, you should probably buy it. Some rental places don't stock the cheap stuff anyway. A good rule of thumb to use to help make the decision is based on how many times you need the tool. One author states that he buys the tool the third time he needs it. Rental stores may sometimes charge as much as one-fourth or one-third of what it would cost if you paid for it. If you need something more than 3 times in a reasonable time period, then it makes sense to buy it.

Rental stores are a good option, but you should use a little common sense. Another option is to ask the building supply store. Sometimes they will rent or loan special tools when you purchase the materials from them.

SEALANTS

Sealants should be used every time you install a threaded pipe, a fitting, or a fixture. They are also used to seal the basket area of sinks and around the edges of fixtures such as bathtubs. There are several, and of course, some are more specialized than others. Some sealants can be used with any type thread while others can be used only on steel pipe threads. A working knowledge of the various sealants will help you do a better job. Figure 4-24 shows a group of these sealants.

Fig. 4-24 *Sealants used in plumbing include Teflon tape, stick-type pipe thread "dope," tube or paste pipe thread dope, plumber's putty, and bathroom (silicone) caulk.*

There are two categories of sealants. The first is used to seal joints between drains and sinks. The second is used to seal threaded connections.

Plumber's putty is a soft, doughy substance. It is usually yellow or tan and is available in cans or plastic tubs. It is inexpensive, and you can buy small containers or large ones. It is used to seal drain connections to sinks and tubs, and to prevent leaks around the bottoms of faucets. Sometimes these areas are sealed with neoprene gaskets, but plumber's putty is still something to keep on hand.

Thread sealants take three different forms: paste, stick, and tape. The paste is now usually purchased in a squeeze tube, but there are still a few around in a can with a swab in the top. The stick form comes in a cardboard tube that can be rubbed across the male threads to fill them with the sealant. Teflon tape is also used to seal threads. It should be wound around the threads in the same direction as the pipe or fitting is turned to tighten it.

A couple of comments about thread sealants are needed here. First, the only thread sealant that can be used on any type of thread is Teflon tape. Most paste and stick sealants can be used only with metallic pipe and should never be used with plastic pipe. Always read the label before you buy; both containers will have this type of information on them. There are some squeeze-tube paste sealants for plastic pipe, but you should read the label before you buy them. All types are relatively inexpensive, and the amateur plumber should keep them in stock.

The last type of sealant used in plumbing is silicone caulk. You use this around showers, sinks, and tubs to form a watertight seal between the wall and the fixture. Don't confuse silicone caulk with painter's caulk, which simply won't work. You can obtain silicone caulk in a variety of colors including white, bronze, gray, black, and clear. You can also buy it in small squeeze tubes or in the tube size that fits your caulking gun.

5
CHAPTER

Working Pipe

T O *WORK PIPE* IS TO USE PIPE TO CONSTRUCT A functional system of some sort. To work pipe effectively, you need to know a lot of things, such as various code requirements, types and sizes of pipe to use, and then the techniques of actually forming and connecting the pipe into a system. The previous chapters covered the theory you need to know to effectively work pipe. This chapter will focus on the knowledge and skills needed to work pipe of several types.

OPTIONS

When you decide to work pipe, there is a reason. You are making repairs, replacing something, or doing new plumbing. You have some options when you begin. If, for example, a pipe bursts during the cold of the night, you may just want to put a temporary patch on the break until later. At some later time, you intend to make a permanent repair. Of course, there are probably a lot of "temporary" patches still in use. Late night or early morning repair jobs tend to be fixed with what you have on hand.

If the job you intend to do is in a spot that's hard to reach and has little room to move your hands or tools, there are special tools and fittings that may help out. You might not ordinarily think of using these, but a little forethought may be of great value. For example, consider both compression and threaded unions for rigid copper pipe. You usually don't use them, but they do provide options for repairs.

Another factor is whether you want to use the same type of pipe. If you are having rust problems with your galvanized pipe, you might want to do your repairs with plastic or copper pipe. Then you need to consider factors such as electrolytic action between two types of metallic pipe, or if you are disconnecting a ground to your electrical system if you use a plastic repair.

What you have on hand also may determine what your options are. Some of us who have muttered at frozen pipes at 5 a.m. or expanded our vocabularies at the difficulty of reaching something have learned to keep a stockpile of parts. There are also a few tools that are handy to have at that moment. It's a good idea to keep a few extra fittings, a compression coupling, a leak clamp (also called a saddle patch), and a few extra pieces of pipe and threaded nipples. Some extra washers or cartridges for your faucets are also wise ideas.

Plumbing components are often between walls, under floors, and in corners that are hard to reach. Sometimes the way to reach a trouble spot is through a narrow opening in a wall or floor. A place that's hard to reach, or very small, is called a *tight spot*. For these areas, special fittings are ideal. Also, you may need short-handled wrenches, extension bars for sockets,

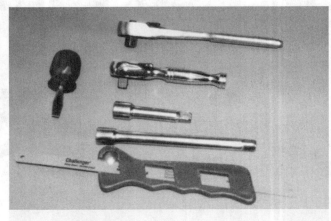

(A)

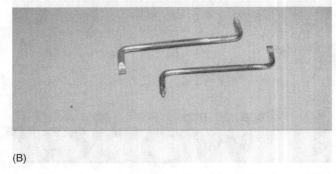

(B)

Fig. 5-1 *Tools for tight spots. (A) A "stub" ratchet handle compared to a regular one. To the left is a stub screwdriver. Underneath are two socket extensions and a mini hacksaw. (B) Offset screwdrivers.*

minisaws, and other tools. See Fig. 5-1. When soldering in these areas with a torch, you may also need to make a heat shield to keep from burning a wall stud or a floor joist. We'll try to mention all these things as we present pipe working details.

PIPE SIZE

Before you begin working pipe, there is another thing you should know about pipe. How it is sized doesn't always seem to make sense. For example, a piece of galvanized steel pipe that's rated as $\frac{1}{2}$-inch pipe has an outside diameter of $\frac{3}{4}$ inch and an inside diameter of $\frac{9}{16}$ inch. None of these dimensions are $\frac{1}{2}$ inch, yet it's rated as $\frac{1}{2}$-inch pipe.

The answer is that pipe is rated by the amount of water that will pass through it. The rated value is obtained by measuring the diameter of the stream of water that comes from the end of the pipe, or the size of the real water flow. Over the years, the standard has become the diameter of the stream of water 1 foot past the end of the pipe at the standard city pressure of 80 pounds per square inch (psi). Over the years, engineers have developed standard water flow tables to easily calculate water supply lines, storage capacities of water tanks, and a lot of other details.

Every kind of pipe has internal friction, even on the flow of water. Rougher pipe walls offer greater friction. Steel pipe has rougher wall than either plastic or copper. For this reason, the actual opening in a pipe is larger than its rated size. However, you buy pipe by its rated size and not by its actual dimensions. To help you understand some of this, Table 5-1 provides this information. Beware, because the sizes shown are only approximate. Remember that there are different wall thicknesses for copper pipe (M, L, and K) and different thicknesses of plastic pipe (STR being a common one). Also, drainpipe thicknesses are different from those of potable water supply pipes.

Table 5-1 Supply Pipe Dimensions

Pipe Type	Rated Size, Inches	Actual ID*, Inches	OD†, Inches	Fitting Depth, Inches
Galvanized steel	1/8	5/16	3/8	1/4
	1/4	3/8	1/2	3/8
	3/8	1/2	5/8	3/8
	1/2	9/16	3/4	1/2
	3/4	13/16	1	9/16
	1	1 1/16	1 1/4	11/16
Copper pipe (type M)	1/2	9/16	5/8	1/2
	3/4	13/16	7/8	3/4
	1	1 1/16	1 3/16	15/16
PVC‡ (STR)	1/2	5/8	7/8	1/2
	3/4	13/16	1 1/8	5/8
	1	1 1/16	1 3/8	3/4
CPVC (STR)	1/2	1/2	5/8	1/2
	3/4	3/4	1	5/8
	1	1	1 3/8	3/4

*ID = inside diameter.
†OD = outside diameter.
‡Dimensions of PVC are different from CPVC dimensions.

The information in Table 5-1 is perhaps the most useful when you have to buy parts and don't know what you have. You can measure the outside dimension of the pipe, read any number or letter codes on it, and use that information to purchase needed parts. When you are in doubt, take a piece of the pipe you are working with you to the store. Because the main sizes you will probably work with are 1/2, 3/4, and 1 inch, you will soon learn to distinguish these sizes.

You might think that the advent of the metric system solved all this. Well, the metric solution was to simply adopt the engineering standards for water flow used in the United States, Canada, and the British Empire. When much of the world was rebuilding after World War II, this was by far easier than converting everything to metric numbers. Since then, some numbers have been converted to true metric systems, but the unified standard (as it was known) is still being used.

WORKING RIGID COPPER PIPE

Rigid copper pipe components are usually soldered together. There are a few pieces that are threaded, but most are soldered. You should refer to Chap. 4 for a list of tools you will need. You should also remember to work safely. If you are doing overhead soldering, a sturdy hat is a good idea. Drops of molten solder can hurt if they land on exposed heads and arms.

The first thing you need to do when assembling copper pipe is to find out how long the pieces should be. Each fitting used takes up some space, and you must allow for it. Figure 5-2 shows how to determine the length you need between the fittings. Next, you should cut the pipe to that length. There is an old axiom about measuring twice and cutting once. It's a good axiom to remember.

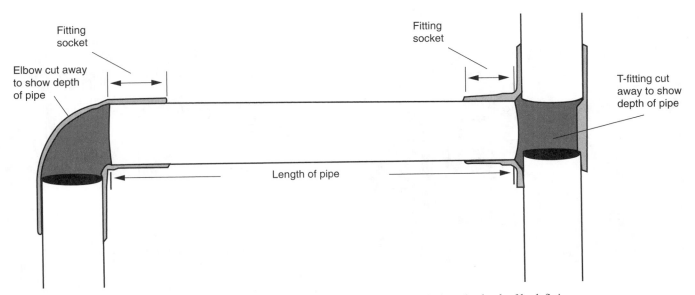

Fig. 5-2 *Determine the length needed for copper pipe by including the depth of both fittings.*

You may use a tubing cutter as in Fig. 5-3. Be sure you ream the burr out of the cut opening to keep the optimal flow, as in Fig. 5-4. It's okay to use a saw on copper pipe. When you use a saw, remember to hold the pipe in a pipe vise, or the pipe jaws of a bench vise, as in Fig. 5-5. You may also cut copper pipe by using a miter box and a hacksaw, as in Fig. 5-6. If you are cutting a piece out of an existing pipe (such as in a leaky place), you may need only a little support from one hand.

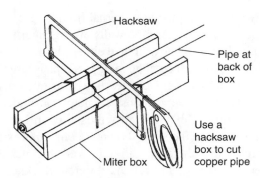

Fig. 5-6 *A hacksaw and miter box can be used to cut copper pipe.*

A mini hacksaw, as in Fig. 5-1, is good for tight spots. A regular hacksaw is handy for working from a vise. One of the authors, however, has found that a reciprocal saw is an ideal saw for working pipe. See Fig. 5-7. The battery-powered reciprocal saw in Fig. 5-8 is the author's favorite. There are no problems with cords or outlets, and a fine-tooth metal cutting blade makes a smooth cut. The battery-powered saw is a bit smaller than its regular counterpart, which makes it easier to handle and better for tight places.

After you have cut a piece to length, make sure you have no burrs on either the inside or the outside of the pipe. See Fig. 5-9. Then check to make sure the length is correct. You have about $1/16$ to $1/8$ inch of leeway.

For best results, it is important to coat both the outside of the pipe and the inside of the fitting with a very

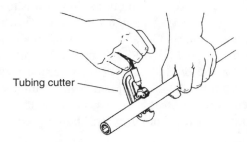

Fig. 5-3 *Using a tubing cutter on copper pipe.*

Fig. 5-4 *Use the point reamer on the side of the cutter to ream the cut opening.*

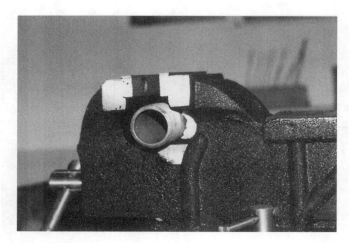

Fig. 5-5 *A good bench vise will have pipe jaws beneath the regular flat jaws.*

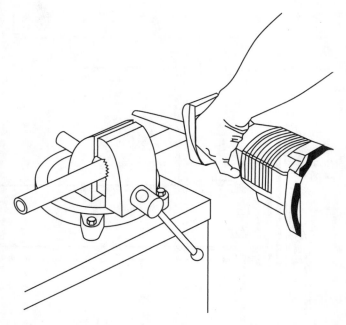

Fig. 5-7 *A reciprocal saw can be used to cut any metallic pipe. (A) Secure pipe in pipe jaws or a pipe vise. (B) Use a metal cutting blade. (C) Put the blade on the line to be cut. (D) Put the front base firmly against the pipe. (E) Press the trigger switch and let the weight of the saw make the cut.*

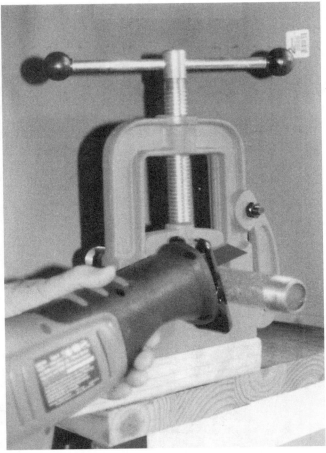

Fig. 5-8 A cordless reciprocal saw is an ideal cutting tool.

Fig. 5-9 A large burring reamer is used on larger pipe sizes. A file is used to smooth the outside.

thin coat of solder before the pieces are joined. This is called *tinning* the joint. If the tinning is thick and lumpy, you cannot slide the pieces together. When the tinning is thick or lumpy, the excess can be wiped off by using a cloth while the solder is still molten. This wiping with a rag leaves a very thin coating of solder on the pipe.

Perhaps the handiest method is to use a special soldering compound, which contains powdered solder in a greaselike mixture of *flux*. Flux is a fluid or paste used to clean the metal when it is being soldered. It is important to use the right type of flux. The flux is simply smeared on both pieces instead of tinning them, and then the pieces are assembled. After assembling either a tinned joint or a compounded joint, the unit must be heated to melt and fuse the solder. More solder can be added, as shown in Fig. 5-10.

Copper pipe also can be soldered without tinning the joints. The solder joints are just coated with flux and assembled. The solder is added to the outside and flows into the joint. This is the easiest method and it works well.

It does take practice and skill to "tin" joints for soldering but it is the best method. When you are starting a new job, it is a good idea to practice with a few short pieces before moving to the actual pipes.

The old systems were soldered using regular solder, which is a mixture of lead and tin. However, lead is a toxic material and could lead to lead poisoning in extreme cases. To prevent lead poisoning, many systems now require a lead-free solder.

Both types of solder are easily obtainable at building supply stores. Either can be used, but the user should be sure to match the correct flux with the solder chosen.

WORKING WITH FLEXIBLE COPPER TUBING

As mentioned earlier, flexible (or soft) copper tubing is purchased in coils of various lengths. Because it is very soft, you can make wide bends by just curving it with your hands. Do not attempt to make sharp bends with your hands. This will cause kinks, which limit the flow of water. You must use a bending tool for sharp bends.

You should measure the route planned for the tubing first. Then purchase a coil that is a few feet longer than you need, just in case. It's also a good idea to have a slight curve at several places to give some slack in the line. This allows for expansion and contraction and for any last-minute adjustments. Next, you should make any required bends in the tubing, using some type of tubing bender. If you refer to Chap. 4, you can see several ideas. Figure 5-11 shows how several tools are used for bending tubing.

You don't use many fittings with copper tubing. You just bend it around corners from the source to the appliance. You need fittings only at each end. However, there are fittings, such as tees, that are available for those cases where you don't wend the tubing directly from one place to another.

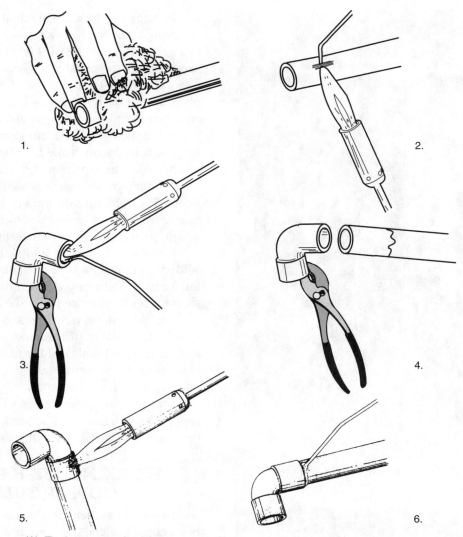

(A). Tinning copper pipe joints
1. Clean end of pipe with steel wool
2. Heat end of pipe and coat with solder
3. Wipe off excess
4. Grip fitting with pliers. Heat and apply solder to joint. Tap out excess
5. Heat fitting and push it onto pipe
6. Heat and apply more solder if needed. Finished joint should be smooth all the way around.

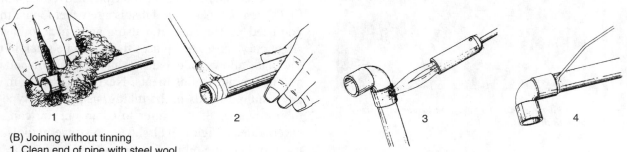

(B) Joining without tinning
1. Clean end of pipe with steel wool
2. Apply flux with brush or squirt tube
3. Push fitting onto pipe and heat
4. Apply solder, melted solder will flow up into joint.

Fig. 5-10 *Two methods of soldering copper pipe: (A) Tinning copper pipe joints. (1) Clean end of pipe with steel wool. (2) Heat end of pipe and coat with solder. (3) Wipe off excess. (4) Grip fitting with pliers. Heat and apply solder to joint. Tap out excess. (5) Heat fitting and push it onto pipe. (6) Heat and apply more solder if needed. Finished joint should be smooth all the way around. (B) Joining without tinning: (1) Clean end pipe with steel wool. (2) Apply flux with brush or squirt tube. (3) Push fitting onto pipe and heat. (4) Apply solder. Melted solder will flow up into the joint.*

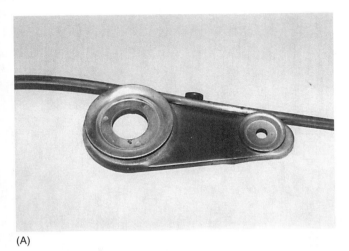

(A)

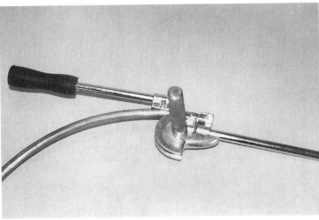

(B)

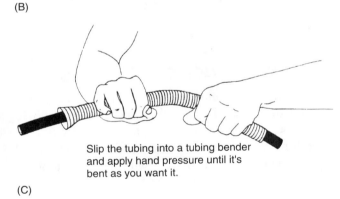

Slip the tubing into a tubing bender and apply hand pressure until it's bent as you want it.

(C)

Fig. 5-11 *(A) Small tubing is bent with a bending jig. (B) Tubing is being bent with a conduit bender. (C) A tubing bender.*

To cut copper tubing, it is best to use a tubing cutter, just as you would on rigid copper pipe. You should ream the opening with either a special reaming tool or the pointed blade on the tubing cutter. Refer to Figs. 5-3 and 5-4.

There are two types of fittings used with soft copper tubing. The first is a compression fitting, as in Fig. 5-12. To use this type of fitting, you slide a compression nut and compression ring past the end of the tubing. Then

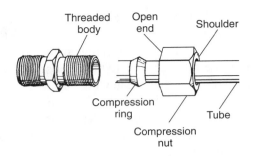

Fig. 5-12 *A compression fitting.*

you insert the end of the tubing into the threaded body of the fitting, slide the compression ring and nut down onto the fitting, and tighten it. The force of the tight compression nut distorts the shape of the compression ring enough to effect a watertight seal. It's a good idea to use Teflon tape or an approved pipe dope on the threads. Remember that the pipe dope for galvanized pipe may not be approved for use on brass, copper, or plastic pipe.

Occasionally you may find a fitting that has a compression ring that is split. That is, the ring has a deliberate cut in it and is not one solid piece. You should never use one of these for water under pressure. They were designed to be used in natural gas systems where the normal pressure is only 4 to 6 pounds per square inch (psi), not the 50 to 80 psi for water systems.

The second type of fitting is considered by many to be more reliable than the compression fitting. Of course, it's also a bit more involved and takes more specialized tools. Figure 5-13 shows how a flared fitting works. To

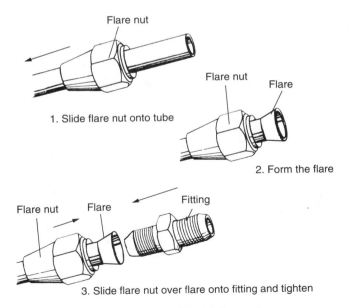

1. Slide flare nut onto tube

2. Form the flare

3. Slide flare nut over flare onto fitting and tighten

Fig. 5-13 *How a flared fitting works.*

make a flared fitting, you need the fitting and a flaring tool, as in Fig. 5-14. First you slide the flare nut down on the tubing with the large opening up toward the joint. Then you insert the tubing into the anvil part of the tool and clamp it in place with the wing nuts. Next position the ram directly over the mouth of the tubing and screw it down tightly. The ram spreads the mouth of the tubing into the correct flare to match the angles on the flare nut and flare fitting.

When you need to replace a curved section of copper tubing, there is an easy way to do this. It is difficult to determine the exact distance from one point to another and to allow for the curve. Instead of fussing with the math, make the replacement bend and allow an extra length on each end of the bend. Then you can place the new piece next to the old one, as in Fig. 5-15, and cut off the ends where you need to.

If, by chance, you must join soft copper tubing to hard copper pipe, you must use special tools called *swages*. These are punches that are inserted into the opening of the soft copper tube and hammered down to form a built-in fitting for the hard copper pipe. See Fig. 5-16. The two pieces are then soldered into place.

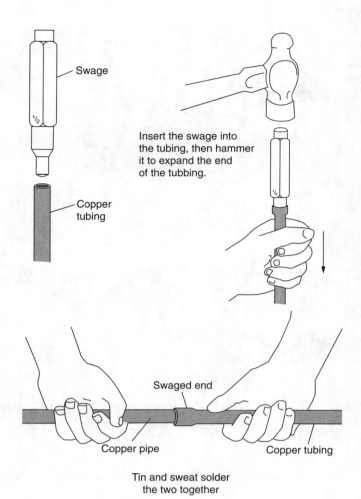

Insert the swage into the tubing, then hammer it to expand the end of the tubing.

Tin and sweat solder the two together

Fig. 5-16 *Using swages to couple copper tubing to copper pipe.*

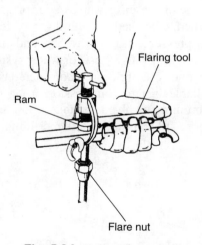

Fig. 5-14 *Using a flaring tool.*

WORKING PLASTIC PIPE

The three most common types of plastic pipe (not flexible tubing) are PVC, CPVC, and ABS. You should remember from an earlier chapter that ABS is only used for DWV. It requires fittings of ABS and special ABS cement. You can work it with the same tools as you use for both PVC and CPVC. When you cement

Fig. 5-15 *An easy way to install a tubing bend.*

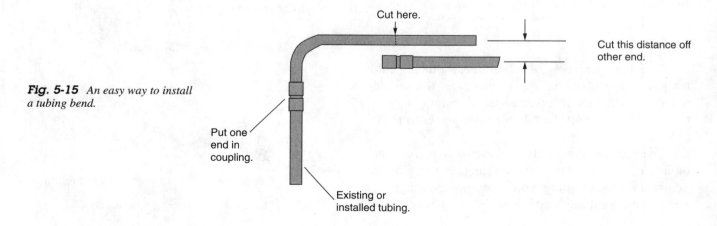

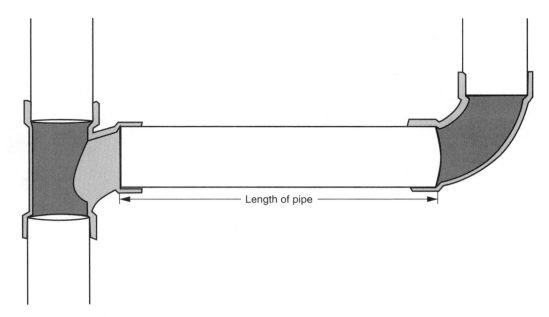

Fig. 5-17 *To cut plastic pipe to the correct length, allow for the distance required to fully seat in the fittings.*

plastic pipe in place, the cement is actually a solvent and the process is really a form of welding.

First, of course, you should measure each piece. Figure 5-17 shows the allowances you should make for the joints and fittings. Next, you can cut plastic pipe with a tubing cutter or a common wood-cutting saw, as in Fig. 5-18. You should also cut the burr off, as in Fig. 5-19.

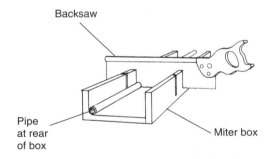

Fig. 5-18 *Use a hacksaw and miter box to cut plastic pipe.*

To join the pipe together, simply coat the inside of the fitting and the outside of the pipe with cement. You do this with the built-in swab in the can, as in Fig. 5-19. Then you push the two pieces together with a slight twisting motion, as shown.

Where you must have an exact alignment of the pieces, you should put them together first without cement. Then mark a visible spot with two lines, as shown in Fig. 5-20. You can take them apart, coat them with cement, and assemble the unit with the marks aligned.

WORKING SOFT PLASTIC TUBING

To work soft plastic tubing, such as PE or PB tubing, special fittings known as *barbed* fittings are used. Refer back to Fig. 3-19. You must allow for the fitting shoulder in flexible tubing, but the shoulder is on the outside rather than internal.

Plastic tubing is easy to cut with a special plastic tubing cutter or with ordinary woodworking tools. PE tubing and PB tubing are not totally limp, and they can be held in a pipe vise or the pipe jaws of a bench vise. You don't need a lot of force holding them, so tighten the vise just enough to firmly hold the work.

You can usually just push the fittings into the opening of the tubing, but it may take some muscle. The larger sizes can be very stubborn. You can dampen the fitting and use a board and a hammer, as shown in Fig. 5-21. Be sure to just tap the board gently. You should also allow the pipe to project from the vise or your hand the length that the fitting will take. Again, see Fig. 5-21. A lot of force can distort the soft fitting and can result in a leak. Persistence and patience work better than strong-arm tactics.

WORKING GALVANIZED PIPE

Galvanized pipe is worked with galvanized fittings. The first step in assembly is to measure the length you need, allowing for the distance the threads travel in each fitting. See Fig. 5-22.

Next, you should select the correct length from an appropriate nipple assortment. Remember that a 4-inch

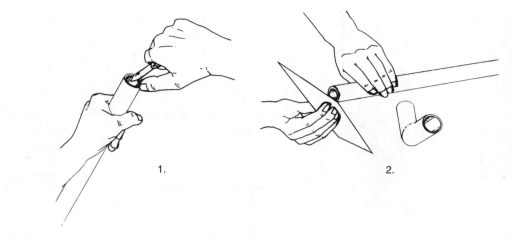

1.

2.

Fig. 5-19 *Welding plastic pipe: (1) Clear inside and outside edges of burrs. (2) Sand glaze off joint area. (3) Coat pipe with solvent. (4) Coat inside of fitting with solvent. (5) Push fitting onto pipe with a slight twist.*

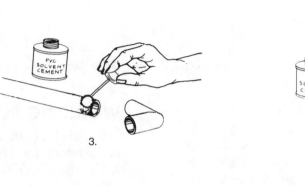

PVC SOLVENT CEMENT

3.

PVC SOLVENT CEMENT

4.

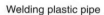

5.

Welding plastic pipe
1. Clear inside and outside edges of burrs
2. Sand glaze off joint area
3. Coat pipe with solvent
4. Coat inside of fitting with solvent
5. Push fitting onto pipe with a slight twist

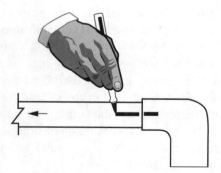

Fig. 5-20 *To keep an exact alignment, assemble the pieces without cement. Adjust the pieces to the correct alignment and mark each piece as shown. Then take them apart, apply cement, and reassemble with mark aligned.*

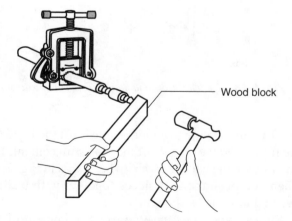

Wood block

Fig. 5-21 *Insert stubborn plastic tubing fittings by gently tapping them in. Use a board as a strike plate to protect the fitting.*

nipple and a 3-inch nipple can be combined to make a 7-inch nipple. If this is not practical, then you must cut the length you need from a longer piece and thread the end. To cut the pipe you can use a pipe cutter, as in Fig.

5-23, or a reciprocal saw or hacksaw. Always remember to cut the burrs from both the inside and the outside, as in Fig. 5-24.

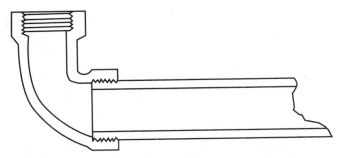

Fig. 5-22 *Galvanized pipe lengths must be cut to include the distance that the threads screw into each fitting. Four or five threads are recommended as a minimum.*

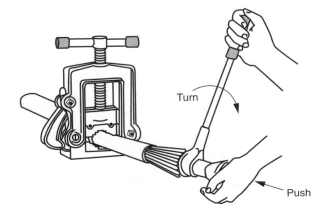

Fig. 5-24 *After you cut pipe, ream out the burr.*

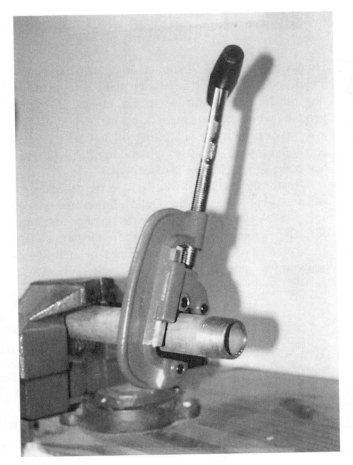

Fig. 5-23 *Cut galvanized pipe to length with a pipe cutter.*

Fig. 5-25 *Galvanized pipe may be held in a pipe vise or a bench vise with pipe jaws.*

To thread the end of the pipe, you must clamp it securely in a pipe vise or a bench vise with pipe jaws. See Fig. 5-25. Next, you select the correct-size threading die, as in Fig. 5-26, and position it in the die wrench, as in Fig. 5-27. The die has sharp teeth that are tapered. This allows you to guide the die onto the pipe and begin with light cuts into the steel. As the die moves onto the pipe, the inner teeth complete the full cut of the thread. It takes some muscle.

There is also a ratchet action for the die wrench. If you look at the die wrench in Fig. 5-27, you will see a black knob on the top of it. The knob can be pulled up and reversed to change the ratchet direction. The ratchet action really comes in handy when you have to thread a pipe in a wall that is hard to reach.

Fig. 5-26 *To thread galvanized pipe, first select the die for that pipe size. The size is stamped on the die face.*

Fig. 5-27 *The pipe die is secured into the die wrench. Note that the wrench has a ratchet.*

Cutting the threads generates heat and metal shavings. You should also reverse the direction and back up about one-fourth of a turn after every two or three turns. This breaks up the chips and allows them to fall from the die. By doing this you will get a cleaner and more uniform set of threads. For best results you should use *cutting oil* that both lubricates the cutting process and removes the generated heat from the teeth of the die. This prolongs the useful life of the die teeth. Old-timers sometimes just used water poured on the threads, but cutting oil will give you a longer-lasting die, better threads, and fewer leaks.

If you are just going to cut one or two threads, you can use your regular oilcan for the job, as in Fig. 5-28.

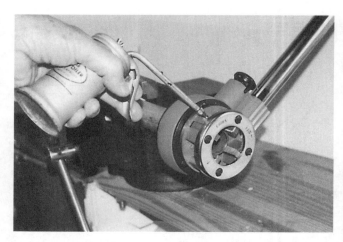

Fig. 5-28 *Cutting oil lets the die cut a better thread.*

However, if you are going to do a lot of this, you should consider either buying or making regular cutting oil. You can purchase special cutting oils from vendors who stock metalworking supplies. You can also make a pretty good cutting oil by mixing equal parts of kerosene and 30-weight motor oil. You can use barbecue lighter fluid as kerosene.

Next, you can begin to assemble the pieces. First you must put a sealant on the threads. Teflon tape, as seen in Fig. 5-29, is easy to use. Just wrap three or four turns of the tape around the threads, and give the tape a sharp tug to break it off. You should wrap the tape in the direction you will turn the pipe, for best results. Figure 5-30

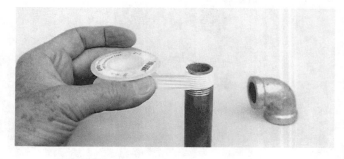

Fig. 5-29 *Using Teflon tape for sealing threads.*

shows how you can use either the stick or the paste sealant. Most experienced plumbers call these sealants *pipe dope*. If you use the paste, squeeze a generous portion on the threads and smear it all the way around the pipe, and back at least four threads. If you use the stick, rub it vertically across the threads, as shown in Fig. 5-30, and continue this all the way around the pipe.

After you have sealed the threads, you may begin screwing the pieces together. You must use two wrenches, as in Fig. 5-31, to do this. The best pipe

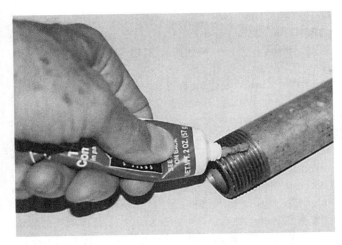

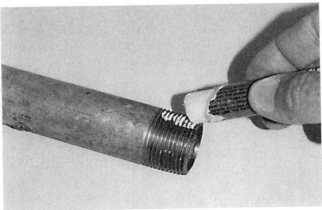

Fig. 5-30 *Apply either paste or stick pipe dope all the way around.*

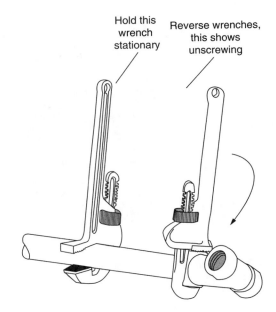

Fig. 5-31 *Two wrenches are needed to work threaded pipe. One is used to hold the work, and the other screws the fitting in place.*

wrenches, also sometimes called *Stillson* wrenches, are made from cast or forged malleable iron or steel. They have hardened steel jaws with sharp teeth that grip the pipe tightly. They are heavy. Some plumbers now prefer wrenches made from forged aluminum with hardened steel jaw inserts.

WORKING CAST IRON PIPE

Working with cast iron pipe may require some special operations and tools. You may also need to be very careful and wear appropriate safety gear. You may be working with molten lead, open fires, and hot metals.

As mentioned earlier, cast iron pipe is heavy and hard to work. Perhaps the most exacting job in working it is cutting it. Before you cut it, however, remember to measure it. Also remember to measure twice and cut once. Make a clearly visible mark all the way around the pipe where you want to cut. Make sure you allow for the distance the pipe will jut into the fitting.

You can use a hacksaw to cut the pipe, but that's a long and tedious job. We recommend that either a reciprocal saw or a chain cutter be used. Both can often be

rented. A chain cutter, as in Fig. 5-32, is wrapped around the pipe, attached, tightened, and then turned just as a smaller pipe cutter. You can turn and tighten the cutter until the piece comes apart, or you can cut about halfway through and break off the end.

If you have someone who can help, you can also use a circular saw with a metal cutting blade. One person

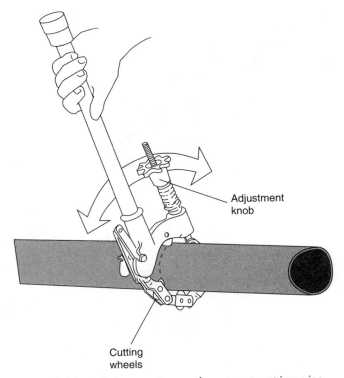

Fig. 5-32 *A chain cutter is a good way to cut cast iron pipe.*

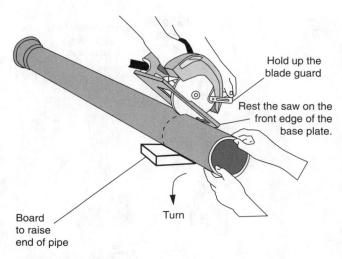

Hold up the blade guard

Rest the saw on the front edge of the base plate.

Board to raise end of pipe

Turn

Fig. 5-33 *A circular saw with a metal cutting blade can be used to cut cast iron pipe.*

holds the saw and makes a shallow cut on the mark while the other person steadies the pipe and rotates it slightly as it is cut. See Fig. 5-33. You should be very careful because the saw can slip, the blade can break, and there will be a lot of hot sparks. Eye protection, gloves, and natural fiber clothing are really vital.

The old way to cut cast iron pipe was to score the intended cut with a hacksaw to a depth of about $1/16$" to $1/8$". This is shown in Fig. 5-34. Then, a cold chisel was used to deepen and widen the cut. Sometimes the end piece just fell off at some point during the chisel operation. If it didn't, then a large hammer was used to give the waste end a series of smart raps all the way around the pipe until the end broke off. This way still works, but it is a lot easier with power tools.

Once the pieces are cut to length, they must be joined. There are some factors you need to consider. If

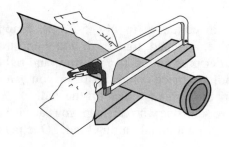

1. Use a hacksaw to make a $1/16$"-deep cut around the pipe.

2. Then deepen the cut with a cold chisel.

Fig. 5-34 *The old way to cut cast iron pipe still works. (1) Use a hacksaw to make a $1/16$-inch-deep cut around the pipe. (2) Then deepen the cut with a cold chisel. (3) Tap the end all the way around.*

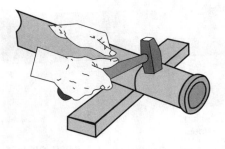

3. Tap the end all the way around.

you are doing a main soil stack, which is a vertical assembly, your codes probably require that you seal each vertical joint with oakum and lead. You will probably be working with whole lengths where hub joints are required. See Fig. 5-35. The hub of the top pipe fits into the opening of the bottom pipe. Oakum is then fitted into the joint and firmly tamped in place. You leave about ¼ to ½ inch between the packed oakum and the top of the flange. Next, lead is melted and poured into the gap between the flange and the side of the pipe. After the lead cools and hardens, it is expanded with a packing iron (sometimes called a *tamping* iron) so that the lead expands and forms an airtight seal against the two pieces of pipe.

Doing horizontal joints is a bit trickier because you simply can't pour the melted lead directly into the joint. You must use a gadget called a *joint runner,* as seen in Fig. 5-36. The joint runner forms a container around the joint with a pouring hole at the top. You pour the melted lead into the hole until it fills up. Let the lead set, and expand it just as with a vertical joint.

Most codes now allow the use of flexible unions on horizontal and sloped runs. Refer to Fig. 3-12. In fact, a variety of neoprene or rubber fittings are made for use with cast iron pipe, and you may be able to use them in your area. These include elbows (els), wyes, and tees. Check with your vendor and permit office first. If you use them, you must cut off the flanges and hubs so that each pipe is just a straight cylinder. In some areas you may be able to buy cast iron pipe without the flanges and hubs.

Each fitting has a flanged opening just like a plastic or copper fitting. You seal the fitting with a clamp around each flange. You must be very careful to support the pipe. The soft fitting will deform and leak if you allow any weight to rest upon it.

WORKING WITH CLAY AND FIBER PIPES

Both clay and fiber pipes are only used for drain or sewer lines outside the home. Most codes require that they be used no closer to the home than 5 feet. These lines must also be sloped properly to allow waste matter to flow into the main sewer. If you put too much slope on the line, the water will run down too quickly and leave the solid matter stuck in the pipe, where it may harden and clog the drain. If you do not put enough slope on the pipe, neither the water nor the solids will drain effectively. This would result in a slow drain and the possibility that the fluids will back up into the house.

Most codes will require either ⅛ or ¼ inch of slope per foot. The most common is probably ⅛ inch per foot. See Fig. 5-37. When you must lay a greater

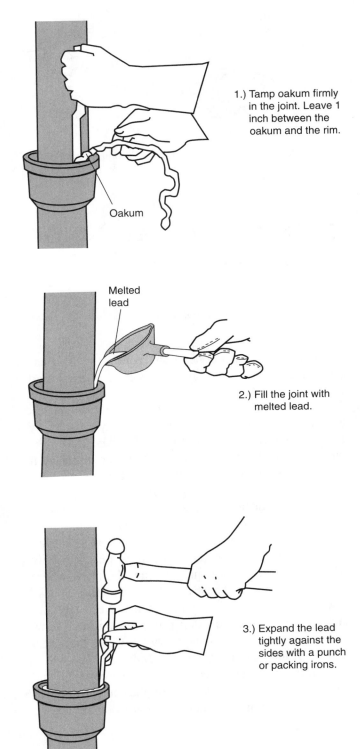

1.) Tamp oakum firmly in the joint. Leave 1 inch between the oakum and the rim.

Oakum

Melted lead

2.) Fill the joint with melted lead.

3.) Expand the lead tightly against the sides with a punch or packing irons.

Fig. 5-35 *Sealing a vertical cast iron pipe joint. (1) Tamp oakum firmly in the joint. Leave 1 inch between the oakum and the rim. (2) Fill the joint with melted lead. (3) Expand the lead tightly against the sides with a punch or packing irons.*

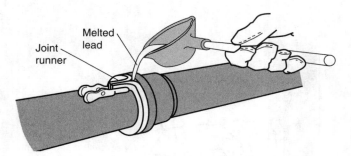

Fig. 5-36 *A joint runner is used to "lead" a horizontal cast iron pipe joint.*

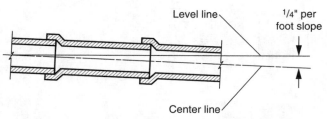

Fig. 5-37 *Clay and other drains should slope ¹/₄ inch per foot.*

angle of slope, most codes now require that you use PVC or steel pipe. The standard inside diameter for all these drains is 4 inches.

Vitrified clay pipe is a ceramic product. The clay has been hardened by partial fusion of the clay particles by firing in a kiln. As a ceramic product, it is hard and brittle, but extremely durable and resistant to rot or corrosion. It is usually purchased in 2-foot lengths that have a wide flange on one end and a straight cylinder on the other.

It is very difficult to cut. Probably the most accurate method is to use a circular power saw, as in Fig. 5-33, except that you must use a blade for cutting masonry rather than metal. You can also use a variety of tools to score (cut a shallow line) the mark where the pipe is to be cut. These could include a file, a hacksaw, a cold chisel, and even a hatchet. Necessity and lack of planning have often produced unusual solutions. However, you always want to cut the extra length off the straight cylindrical end.

Once the line is scored, you can tap around the line with a hammer to break the piece off the main pipe. It will usually break up in pieces rather than as a whole piece.

Once you have all the pieces formed, you must lay the pipe in the trench at the proper slope and then cement the joints. Ordinary masonry cement will work. It is often very tedious and time-consuming to get all the short lengths of pipe sloped properly.

Bituminous fiber pipe is a pipe made from a rot-resistant fiber, such as fiberglass, impregnated with bitumen, a form of tar. It is a lot easier to work with than clay pipe. Fiber pipe is very light in weight compared to clay pipe, but perhaps the best advantage is that it comes in lengths of 8 or 10 feet. The longer length makes it a lot easier to lay at the proper slope.

Both ends of the pipe are tapered so that special fittings, as in Fig. 5-38, may be used to join the pieces. These fittings are sealed with a tar-based cement. Fiber pipe can be cut with ordinary wood-cutting tools such as a handsaw or a power saw with a regular wood-cutting blade.

When you have to cut fiber pipe to a shorter length, probably the best and easiest way to join them is with a flexible sleeve coupling. You must cut the taper off the next pipe to get a good seal, though.

OTHER PIPE TYPES

There are a few other types of pipe that we haven't mentioned so far. These include brass pipe and chrome-plated brass or copper pipe. Brass pipe can be threaded or soldered. You work it just as you would copper pipe if you solder it, or just as galvanized pipe if you thread it.

Chrome-plated pipe shouldn't be soldered because the heat can discolor or damage the plating. Any type of plated pipe is normally worked with compression fittings, just like working soft copper tubing. In most cases, the fittings will also be chrome-plated and sized to fit the pipe you are working.

There are a number of plastic fittings available that can be used with soft copper and with various types of plastic lines. These are usually PB types of plastics with compression fittings. The advantage of plastic fittings is that they are softer and deformed to make a seal with less force.

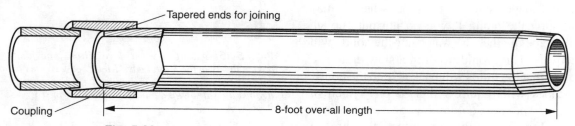

Fig. 5-38 *A joint runner is used to "lead" a horizontal cast iron pipe joint.*

6
CHAPTER

Repairing and Replacing Plumbing

IN PREVIOUS CHAPTERS, WE HAVE PRESENTED details on codes, materials, tools, and pipe working procedures. A homeowner, however, probably has to fix problems with existing plumbing more than anything else. Most of these things are minor problems such as fixing a drippy faucet, a clogged drain, or a problem toilet. However, you may also be faced with a broken or damaged pipe. You might also want to add minor improvements such as cut-off valves for various fixtures. This chapter will focus on these problems and will start with the most common problem and work toward the more difficult ones.

Before you make a trip to the store, it is a good idea to know the brand name and model type of the item. The brand name is usually pretty easy to find. It is often on the back of faucets, but is usually visible. Model numbers are difficult, and you may not be able to find them. If that is the case, then it is a good idea to take the part itself to the store. This way you can find an exact match.

FIXING A LEAKY FAUCET

There are a lot of different types and locations of faucets. A faucet is just a valve used to turn the water on or off. Faucets are used in bathroom lavatories, tubs, and showers. You find them around sinks for kitchens, laundries, and bars. You use them outdoors to water the yard, and you may find them as hookup sites for clothes washers. With extended use, they will all develop leaks.

Most faucets are variations of a globe valve (refer to Chap. 3). Globe valves depend upon a simple washer to seal off water flow, and most are relatively easy to repair.

Faucet repair can usually be done with common tools, but you may need a couple of extra things, as mentioned in Chap. 4. These could include some penetrating lubricant, a handle puller, a facing tool, and a square wrench. For some tub/shower applications you may also need deep-well sockets and other tools.

As you might expect, recent years have complicated the simplicity of faucet repair with a variety of cartridge-type valves. Figure 6-1 shows a couple of these gadgets. Almost all the cartridge systems use a combination of several O-rings to make the seals. Unfortunately for the amateur plumber, they all seem to be specially made and odd-sized. You can't find new O-rings that will work. You don't repair cartridges; you must replace them. While they can simplify the repair process, they are more expensive.

Cartridge systems often use different cartridges for hot and cold faucets. Some manufacturers will

Fig. 6-1 *Cartridge-type faucet valves may be used in lavatory, bath, and kitchen faucets. They are often color-coded for hot or cold faucets.*

color-code these (usually red for hot and blue for cold), while others may use numbers or letter codes.

If you are going to replace kitchen and bath faucets, find a brand that uses the same type of cartridge for all the different faucets. You can find systems for kitchen sinks, bathroom lavatories, and tubs that have interchangeable cartridges. If you only have to use one type of cartridge, you can keep a couple in your stockpile and fix any of the faucets with what you have. It also simplifies the buying process when you need new cartridges.

Replacing faucet cartridges The first step in replacing a cartridge is to turn the water off. This can usually be done at a water supply valve underneath the fixture, as in Fig. 6-2. Then take the faucet handle off. You may not even know your system uses cartridges until you

Fig. 6-2 *The first step in faucet repair is to turn off the water at the supply valves underneath.*

do this. If you have a cartridge system, you will see a metal or plastic nut on top of the plastic cartridge.

If the handle is stubborn, you may need to squirt a penetrating lubricant under the handle. Give it a few minutes, and then try to pry one side gently with a screwdriver. If the faucet handle moves up a bit, then pry from the opposite side and so on until you can remove the handle. If this doesn't produce results, you may need to use a handle puller, as in Fig. 6-3.

Once you have exposed the retainer nut, you can unscrew it with a wrench or channel lock pliers. You can often pull out the cartridge with your fingers. Just grasp the stem and give it a tug. If this doesn't work, then use a pair of pliers to pull on the stem. See Fig. 6-4.

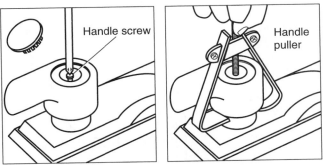

Fig. 6-3 *Remove faucet handles. Stubborn handles can be removed with a puller.*

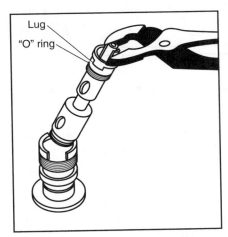

Fig. 6-4 *Pull the cartridge out with pliers. Some have lugs and notches for correct alignment.*

Sometimes you won't get the whole cartridge out. Only the inner part will come out with the stem. If this happens, it's no big deal. Usually, once the inner core is out, the outer core will come out easily; but if it doesn't, just grasp some part of the outer core and tug some more. You can even pry it out with a screwdriver if you must. If you break off part of it, just keep working

on the rest until it comes out. You're going to throw the old cartridge away, so don't worry about damaging it.

Align the notch on the faucet housing with the tab on the cartridge before you insert the new cartridge. Then just push it down firmly. Some types suggest that you coat the new cartridge with waterproof grease before you insert it, and others don't. If the directions with the new cartridges don't suggest it, don't. Then you screw the retaining nut back on, replace the handle, turn on the water supply, and it's done.

Replacing a faucet washer   Replacing washers to fix a leaky faucet is still a common job. Washer-type faucets are very common in older kitchen and bathroom systems, and they are still used in some of the more expensive systems. They are still the most common types used as hose bibs for outdoor use. Technically, they are globe valves, but you may encounter other terms. Probably the most common term is compression faucet. It is so called because the coarse screw of the stem compresses the washer against the valve seat to stop the flow of water.

Figure 6-5 shows a typical kitchen or bath faucet that uses a washer to effect a seal. We won't present separate sections on changing washers in kitchens, sinks, or tubs, because they are really pretty much the same. There is a stem to which the washer is attached and usually a seal around the stem. You may get a leak

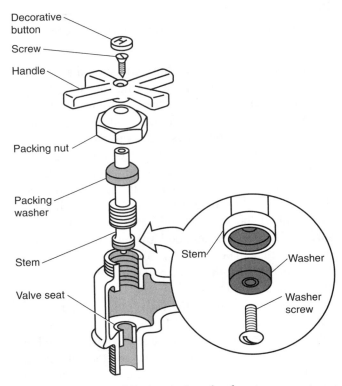

Fig. 6-5 *A typical washer faucet.*

at either place or both places, and both are relatively simple to fix.

As in cartridge replacement, the first two steps are to turn off the water at the supply valve and remove the handle. Next, there will likely be a nut that is both a seal around the stem and a retaining nut. You remove this nut by unscrewing it with a wrench or channel lock pliers. If it is stubborn, give it a shot of penetrating lubricant and a few minutes to work. Then unscrew it. You can unscrew the stem as in Fig. 6-6. Once you have the stem out of the faucet body, unscrew the retainer screw at the end of the stem and remove the old washer. A bad washer will have a deep circular dent or crease in it. To complete the job, you attach the new washer and reverse the process to assemble the faucet.

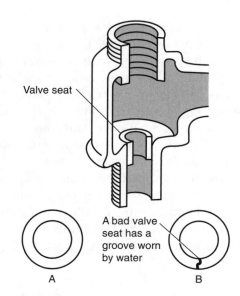

Fig. 6-7 (A) A good valve seat and (B) a bad valve seat.

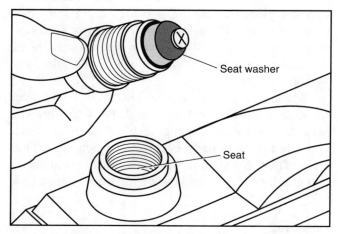

Fig. 6-6 Unscrew the stem and inspect the washer.

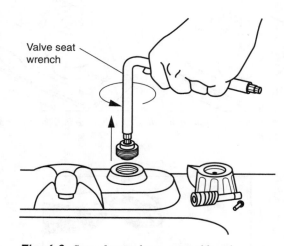

Fig. 6-8 Some faucets have removable valve seats.

However, there is one thing you should always do when you take the stem and washer out of a faucet. Sometimes the leak will not be due to a bad or worn washer. Leaks occur when the seat of the washer within the faucet body goes bad. This can be the result of a flawed casting or the erosive effect of moving water. Either way, a very small fault in the seat can result in a very persistent leak. Look inside the faucet and locate the seat, as in Fig. 6-7. If you can't see it well, use a flashlight. If the seat is good, it will present an even surface. If the seat is not good, there will be a dark line or gap in the surface. If you can't tell, rub the tip of a screwdriver over the surface of the seat. If you feel a bump, there is a fault.

If you find a fault in the valve seat, check to see if you can unscrew the seat. If the hole in the seat is square, you probably can. Use a seat wrench, as in Fig. 6-8, and unscrew the seat. Then lay the face of the seat on a file or a piece of sandpaper, and rub the seat back and forth. Check the face of the seat after every two or

three strokes. Once you have an even, bright face with no lines, the seat can be replaced.

Most faucets do not have a removable seat. The valve seat is molded into the faucet. If your faucet has this feature, use a valve seat tool, as in Fig. 6-9, to resurface the seat. It is probably best for the beginner to do this by hand. More experienced plumbers could use the tool with a power drill at a very low speed. At any rate, all you want to do is to remove enough material to leave a flat surface with no bump or line in it.

Some washers are flat, and others are conical. See Fig. 6-10. Flat washers are easier to work with and give you some wiggle room you don't have with a conical washer. If, for any reason, you don't have a washer that will work (or it's late at night and no stores are open) you can make a temporary repair by just turning a flat washer over and using the unused side. This isn't

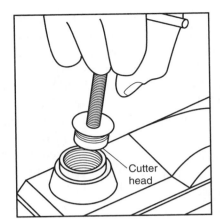

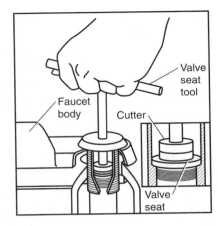

Fig. 6-9 *A valve seat tool is used to resurface a valve seat.*

Valve seat tool

Faucet body

Cutter

Cutter head

Valve seat

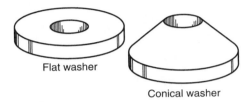

Flat washer

Conical washer

Fig. 6-10 *Some washers are flat and others are conical.*

considered a permanent repair, as the washer will have lost much of its needed resilience. However, some temporary repairs last a long time. Don't worry, your faucet will let you know when a new washer is needed.

As you reassemble the faucet, check the stem for mineral stains. If stains are present, you probably have some seepage around the stem. You can usually stop this seepage by just tightening the cap a quarter turn or so. But before you put it all back together, check to see if the cap washer or packing gland is still in good shape. If it is a felt washer, you may need to add grease or oil. If it is damaged, you should replace it. If you don't have, or can't find, a washer for this, you may use Teflon string, as in Fig. 6-11.

Either way, clean the stem with steel wool or very fine sandpaper. Reassemble the faucet and tighten the cap securely. Put the handle back on, but don't fasten it. The handle should turn fairly easily, but you should be able to feel some resistance. Next, take the handle back off and turn the water supply back on, and look for leaks. If you have some seepage around the stem, tighten it just a little more. You should tighten the cap just a little past the point where the seepage stops.

One-Handle Faucets

In recent years, some faucets feature a single handle, as in Fig. 6-12, that controls hot and cold mixing and volume. Some have a push/pull volume adjustment that turns to either side to adjust the mix of hot and cold water. Others have only an on/off adjustment, but turn to either side to adjust the mix of hot and cold

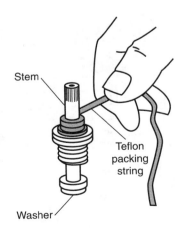

Stem

Teflon packing string

Washer

Fig. 6-11 *A leaky handle can be fixed with a new packing washer or with Teflon packing string.*

Fig. 6-12 *A typical one-handle faucet.*

water. There are versions used in kitchens, bathroom tubs and showers, and lavatories. There are a variety of mechanisms, but there are similarities. One-handle systems usually have either a ball valve mechanism or a cartridge variation of the ball valve. There is one other system frequently found. It is a *disk* system used by Reliant or American Standard. Each of these three systems will be covered.

Base leaks   Kitchen faucets also have another source of leaks. Most kitchen faucets feature a spout that swivels from side to side. Leaks around the base of the spout are caused by worn or uncompressed O-rings (Fig. 6-13). If this is the only leak, try tightening the rounded cap. To keep from damaging the cap, wrap two or three turns of electrical tape or masking tape around the jaws of your pliers, as in Fig. 6-14B. If tightening the cap does not eliminate the leak, then you must replace the O-rings. There are usually two, one near the top and one near the bottom.

If the faucet is leaking from the spout, this means that the seals in the valve itself are bad. To fix this, you must replace the cartridge or the separate seals.

Cartridge faucets   Cartridges for one-handle faucets are actually versions of a two-gate ball valve. There is one opening for the hot water and one for the cold water. The stem of the valve or cartridge is turned one way or another to present the valve openings to the supply openings. The first step, of course, is to turn off both the hot and cold water supply valves.

To replace a cartridge, shown in Fig. 6-14A, first pry off the cover cap and unscrew the handle from the valve stem (Fig. 6-14B). Remember that you should have turned the water off for both the hot and cold water supply lines.

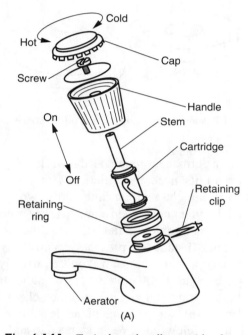

Fig. 6-14A   *Typical one-handle cartridge faucet.*

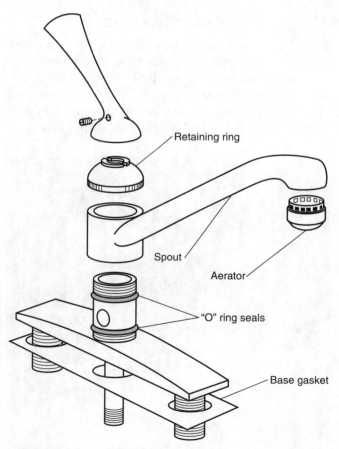

Fig. 6-13 *Leaks at spout bases can sometimes be fixed by tightening the retaining ring.*

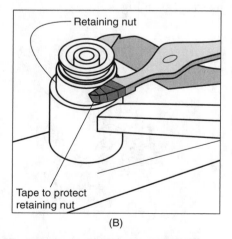

Fig. 6-14B *Remove the handle and unscrew the retaining ring. Cover the plier jaws with tape.*

Next, use channel lock pliers to unscrew the plastic retaining ring. You may find another retaining clip as well. If so, remove this clip (Fig. 6-14C). Then grasp the stem of the cartridge with your pliers and pull it out of the housing (Fig. 6-14D). If it does not all come out, remove the rest as well.

There is only one thing to really check before you insert the new cartridge. Most cartridge systems have a notch or "key" of some sort to ensure that the cartridge is inserted correctly. Refer to Fig. 6-14. Line up the notch with the tab on the cartridge. Then push the new cartridge firmly into place. You can reverse the process for assembly.

It is a good idea to replace the two O-ring seals on the spout while you have the unit apart. With the retaining ring off, pull straight up to remove the spout. To remove the old O-rings, you can either pry them off with a small screwdriver or simply cut them with a knife. Gently roll the new O-rings into place and replace the spout.

Ball faucets Ball faucets feature a plastic ball with two openings. These openings match the two water entry ports, one for hot water and one for cold water. Each of the entry ports inside the valve housing will have a spring and a valve seal. The valve seals are usually made of neoprene, and with use they wear away. The first step, of course, is to turn off both the hot and cold water supply valves.

To repair a ball faucet, first remove the setscrew with an Allen wrench, as in Fig. 6-15. Remove the handle and expose the cap. Wrap your pliers with tape and unscrew the cap. Then remove the cam, washer, and ball. Check the ball for wear or damage. Next, use needle-nose pliers to reach into the body of the faucet and remove the valve seals and springs. There are two of each, one for hot water and one for cold water.

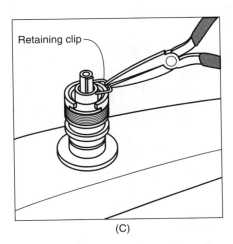

Fig. 6-14C   *Remove the retaining clip.*

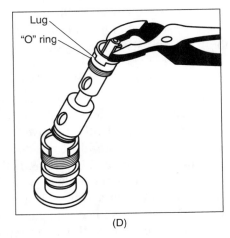

Fig. 6-14D   *Pull the cartridge out with pliers.*

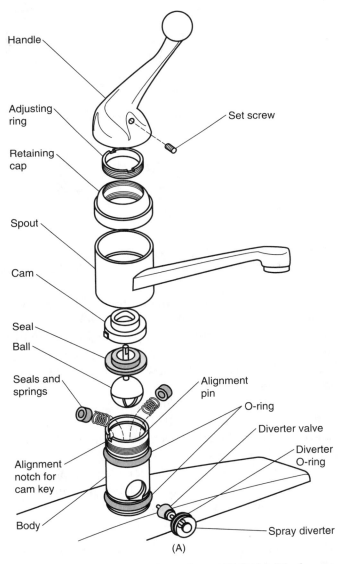

Fig. 6-15   *Repairing a ball-type faucet. (A) Parts of the faucet.*

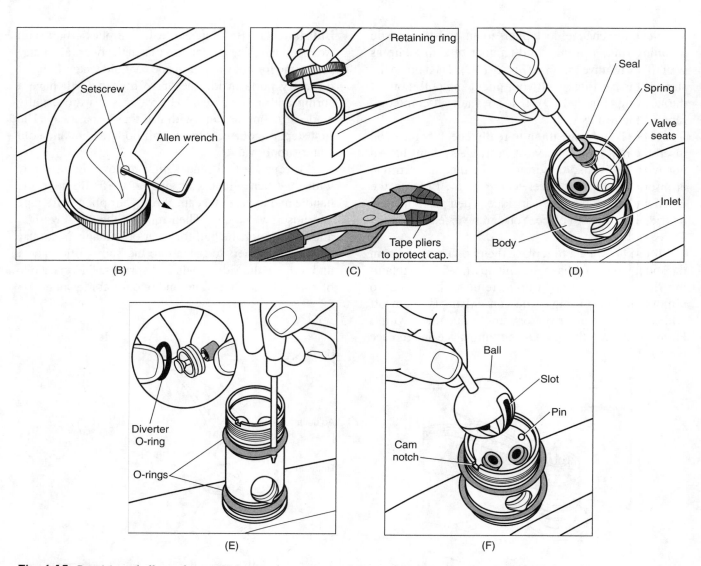

Fig. 6-15 *Repairing a ball-type faucet. (B) Loosen the setscrew and remove the handle. (C) Unscrew the retaining ring with taped pliers. (D) Remove and replace old seals and springs. (E) Replace the O-rings on the body. (F) Reassemble.*

Because it is not as common for the ball to have damage, most stores will sell kits with just the seals and springs. They will carry separate ball repair kits as well. In some cases you may have to purchase a kit with both seals and a new ball if separate units are not sold.

To complete your repair, insert the springs and valve seals. Then insert the ball and cap it with the cam and cam washer. Finally, attach the handle and fasten it in place with the setscrew.

Remember that it is still a good idea to replace the O-rings for the spout while you have the unit apart. Use the same procedure as described for the cartridge system.

Disk faucet Fixing a disk usually means taking it apart and removing the lime or mineral deposits. See Fig. 6-16. The very first step is to shut off both the hot and the cold water supply valves. Then take off the handle. You may encounter either a screw under the cap or a setscrew holding the handle in place. Check the handle to see which system is used, and remove the handle.

Unscrew the cap to expose the cylinder, or disk. The disk is usually held in place with three setscrews. Unscrew the setscrews and remove the disk. On the bottom are three seals. Gently remove these and check the disk. Be very careful of the disk, as it is usually a ceramic material and is fragile. If there are lime or mineral deposits, remove these with a plastic scouring pad. Do not use abrasives such as sandpaper or steel wool. Once the disk is clean, check the disk for cracks or scratches. You will need to replace the disk and seals if the disk is cracked or scratched. If

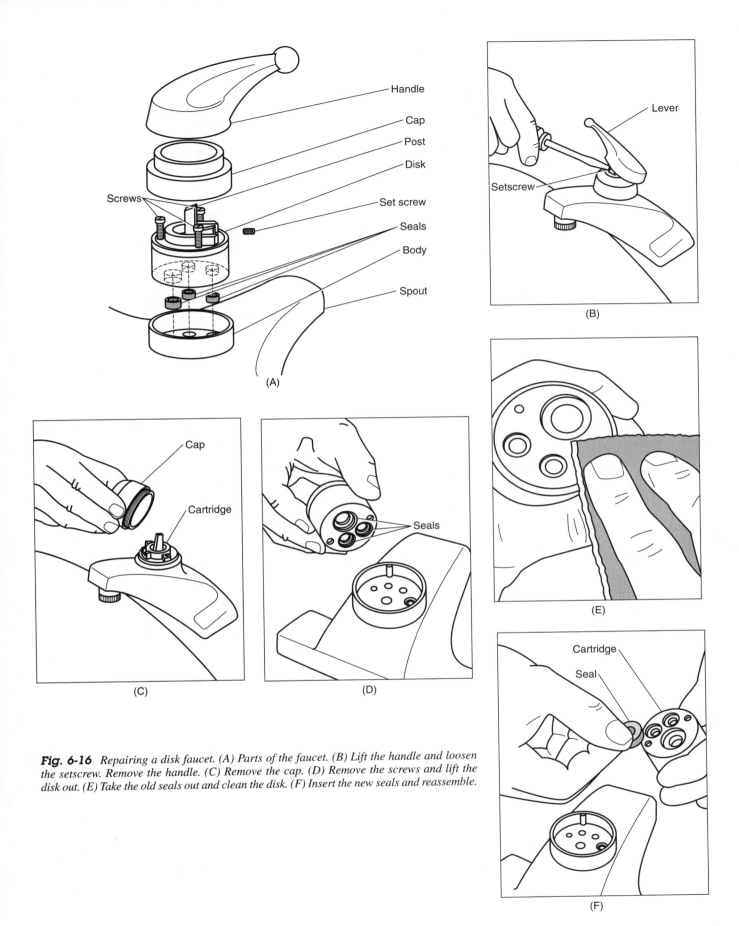

Fig. 6-16 *Repairing a disk faucet. (A) Parts of the faucet. (B) Lift the handle and loosen the setscrew. Remove the handle. (C) Remove the cap. (D) Remove the screws and lift the disk out. (E) Take the old seals out and clean the disk. (F) Insert the new seals and reassemble.*

the disk is not cracked, you probably only need to clean it.

Reinsert the valve seals and reassemble the unit. Before you turn on the water supply valves, move the faucet handle to an on position. This prevents the surge of water or air from damaging the ceramic disk. Check the unit for leaks and drips. If the drip persists, you will need to replace the disk.

Tub and Shower Faucets

Tub and shower faucets are usually either globe valve/washer types or cartridge types. The mechanics are the same or very similar as on the valves used in sinks and lavatories. However, there are some variations in the faucets for these uses. First, the external fittings around these should be sealed from splash water. This can be done with bathroom caulk around the edges or by filling the mountings with plumber's putty and removing the excess after the fitting is in place. See Fig. 6-17.

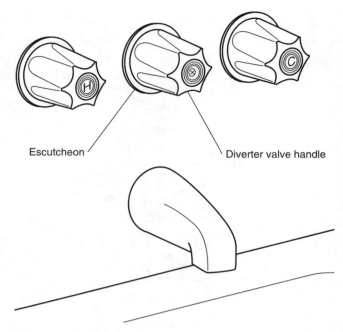

Fig. 6-18 *A three-handle tub/shower system. The middle handle is the diverter valve to switch from tub to shower.*

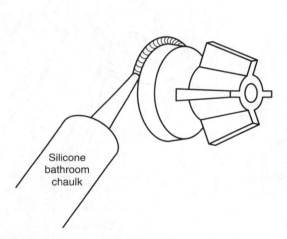

Fig. 6-17 *Faucet caps (escutcheons) in tubs and showers should be sealed from splashes.*

If you have a tub/shower combination, then there is a special valve system to divert the water from the tub spout to the shower. The valve is called a *diverter.* There are separate diverters for each of the systems involved.

If you have a three-handle system, you have a hot and cold handle and a diverter handle. This controls the diverter valve, as in Fig. 6-18. If you have either a two-handle or a one-handle system, then the diverter is located in the tub spout. See Figs. 6-19 and 6-20. These diverters in the tub spouts operate only after the water is running. Some newer types may have a ring

on the outlet that is pulled down, while the standard versions use a lever or rod that is pulled up.

Spout replacement You usually can't repair a leaky spout. The leak probably comes from one of the valves. If the spout leaks a bit during a shower, that's normal, so don't worry about it. Replace a spout when the diverter is bad or when the lime and mineral deposits have ruined it in some way. Before you replace it for lime deposits, try some lime removers. They can do wonders.

However, if you do decide to replace the spout, it is a relatively easy procedure. First, check the bottom for a setscrew. If it has one, remove it. See Fig. 6-21. Then, as shown, stick the end of a large screwdriver or dowel into the spout and unscrew it. If this doesn't work, bring out the heavy artillery and use a big pipe wrench. If you are going to replace the spout, you needn't worry about marring its appearance.

This should expose the supply pipe. Sometimes, when you unscrew the spout, the supply line also unscrews. No problem. Either unscrew the pipe from the spout, or buy a new pipe nipple the length of the old one.

To install a new spout (be sure you got one with a diverter on it, or your shower won't work), put some pipe dope or Teflon tape on the supply pipe thread and screw it back on. If you replace the supply pipe, be sure to use a sealant on the threads into the fitting. Be sure to use a sealant around the edges of the spout where it touches the wall. Either plumber's putty or bathroom caulk will work.

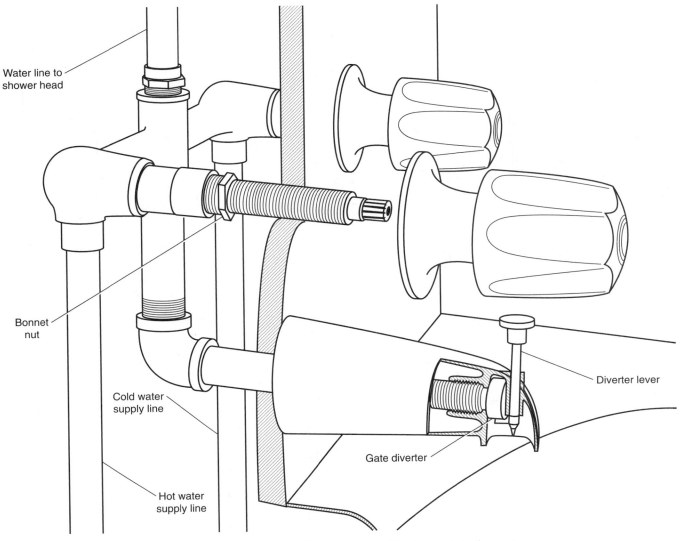

Water line to
shower head

Bonnet
nut

Cold water
supply line

Hot water
supply line

Diverter lever

Gate diverter

Fig. 6-19 *A two-handle tub/shower system. The diverter is in the spout.*

Two- and Three-Handle Faucet Repairs

The two-handle and three-handle faucets are very similar. Each valve is constructed in a similar manner, and each may use either a washer or a cartridge. Also, each will have a cover of some type to provide a pleasing appearance and to seal the interior of the wall from splashes. Refer to Fig. 6-17. The replacement of washers or cartridges is about the same as mentioned earlier. For bath/tub combinations, they are in the wall and are sealed differently.

Note that the cover, officially named an *escutcheon*, is usually fastened in place with a setscrew. The handle is usually kept in place with a stem screw, which may be under a cover or exposed.

And, of course, before you tear into any of the parts, you should turn off the water. Many tubs and showers do not have supply valves, so you may need to turn off the water at the main supply valve near the meter. Some homes have valves that allow you to turn off the water to just a room or two while water to the rest of the house remains on. Find out what you need to do to shut off the water before you start.

Once you have turned off the water, remove the handle and the cover to expose a hole in the wall. The valve body may be visible, or it may be recessed a good bit. If you have a metal valve stem, you probably have a washer system. If you have a plastic valve stem (or a metal stem inside a plastic cylinder), you probably have a cartridge system.

Then remove the *bonnet* or retaining nut, as in Fig. 6-22. If it is above the surface of the wall, you can use a regular wrench. If it is below the surface, you usually need a deep-well socket. Some bonnets are large enough

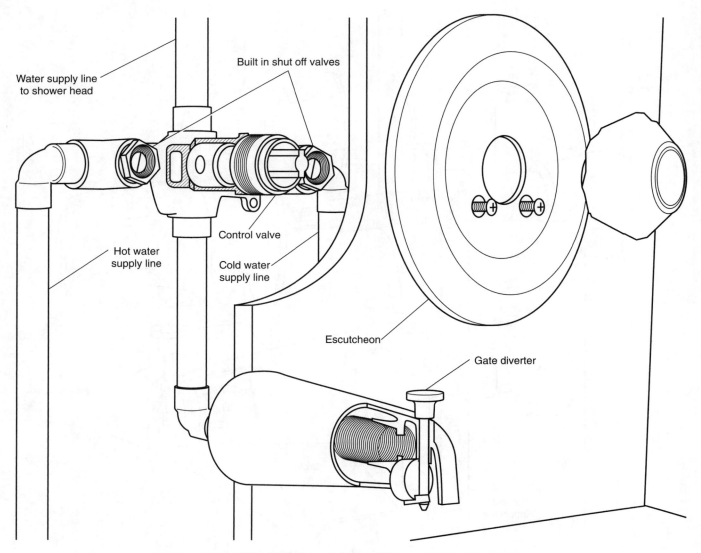

Fig. 6-20 *A one-handle tub/shower system.*

that a standard socket set may not be large enough. Sizes of deep-well sockets to have in your toolbox are ⁵/₈, ³/₄, and 1 inch.

Washer replacement   The bonnet on a washer system, as seen in Fig. 6-23, may also control leaks around the valve stem, so check the stem for mineral deposits that would indicate seepage. Seepage is controlled by washer (or packing gland) compression around the stem. If there is seepage, you just tighten the bonnet a little until it stops. As you take the unit apart, check the packing in the bonnet. If you have had seepage, you may need to replace the packing when you reassemble the unit.

Once you have the bonnet removed, use a deep-well socket to remove the valve. You may have one or more O-rings on the unit. Normally, you do not replace these. The washer is attached to the bottom of the stem with a screw. Remove the screw, and remove the washer. A bad washer will usually have a deep groove in it, and it may also have tears or cracks. Just as with regular washer replacement, you can often turn the old one over and reuse it in a pinch. A new washer is the best choice, though.

You reassemble the unit in reverse order. Screw the bonnet on tightly enough that there is pressure on the stem, but make sure you can still turn it with the handle.

When you replace the cover, or escutcheon, seal it to the wall with plumber's putty or bathroom caulk.

Cartridge replacement   The bonnet on a cartridge system is usually the retaining nut. Once you have the

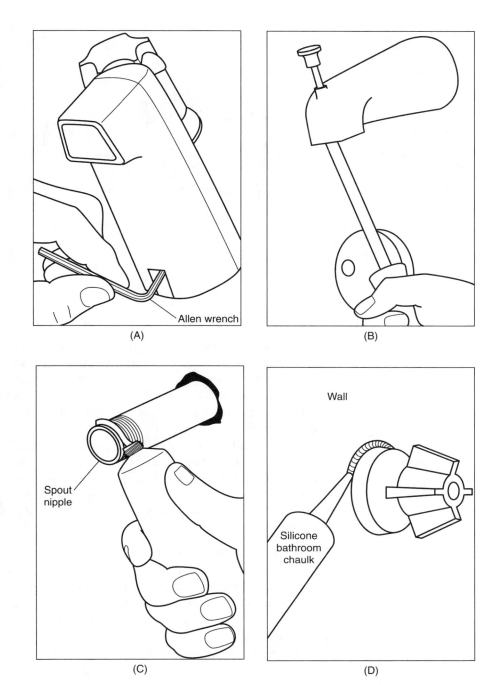

(A)

(B)

Allen wrench

Spout
nipple

Wall

Silicone
bathroom
chaulk

(C)

(D)

Fig. 6-21 *To replace a tub spout: (A) Check underneath for a setscrew. If there, remove it. (B) Use a tool or pipe wrench and unscrew the spout. (C) Put tape or pipe dope on the threads and screw the new spout on. (D) Seal around the spout.*

bonnet off, you simply pull the cartridge out. See Fig. 6-24. Replace the cartridge. Remember that there will be a notch on the housing and a lug on the cartridge to align.

When you reattach the cover, or escutcheon, use a sealant to prevent water from splashing into the wall.

One-handle faucet repair One-handle tub/shower faucets have a single valve that controls the flow of the hot and cold water and the volume. It's called a control valve. They are very much like one-handle valves used in sinks and lavatories. There are three major differences, however. The first is in access and position rather than mechanical makeup. The second is that the mechanism has built-in shutoff valves for the hot and cold water supply lines. The third major difference is that the cartridge is repairable and is fixed with one or two O-rings. Refer to Fig. 6-20.

To repair a one-handle control valve, first you remove the handle and cover plate, or escutcheon. It is

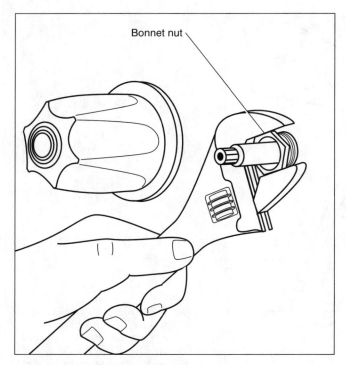

Fig. 6-22 *Unscrew the bonnet nut on the shower faucet. You may have to chip away some tile or mortar. If the nut is recessed, use a socket wrench.*

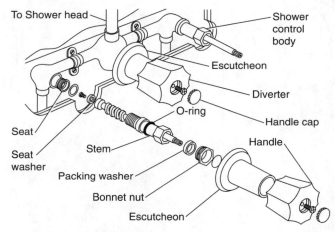

Fig. 6-23 *The bonnet and packing washer can be tightened to stop stem leaks.*

usually held in place by one or two screws. Removing the cover will expose a fairly large opening in the wall. As in Fig. 6-20, you will see a central valve with screws on either side of it.

These screws are the supply line cutoff valves with the hot water supply normally on the left. The cutoff valves are screwed all the way in to cut off the supply and all the way out to turn it back on. So naturally, the next step is to cut off the water by screwing both valves all the way in.

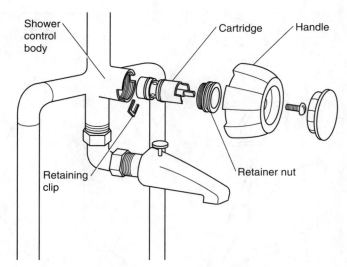

Fig. 6-24 *A cartridge tub/shower faucet.*

Once the water supplies are off, remove the bonnet or retaining nut on the central control valve. Then grab the stem with pliers and gently tug the valve out of the valve seat or housing. You should find one or two O-rings near the end of the cartridge. These are what you remove and replace.

To reassemble the unit, reverse the process. Replace the cartridge and the bonnet, and then turn the supply valves back on. When you replace the cover, be sure to use a sealant around the edges. This will prevent water from splashing back into the wall. Then replace the handle.

Diverter valve repair Diverter valves on three-handle systems are usually globe (compression) valves and can be repaired by replacing a washer. About the only time you need to do this is when the water flow from the showerhead is very low and most of the water runs out the spout.

The repair procedure for a globe-type diverter valve is the same for any washer-type faucet. When you reassemble the valve, be sure you seal the cover against the wall.

Diverters for one- and two-handle systems are in the spout and are usually not repairable. To fix one, you simply replace the spout as previously described.

Other Faucet Problems

Leaks are not the only problems that occur with faucets. Sometimes the water flow from the spout seems inadequate. If there is a sprayer on your kitchen sink, the water flow from it may be weak. And sometimes you continue to have a leak even after you have "fixed" a faucet.

The most common cause of low-flow problems in both lavatory and kitchen faucets is a clogged strainer/aerator. It is commonly called an *aerator,* and its main function is to reduce splashes from a solid stream of water. By mixing air with the water, splashing is reduced.

The aerator is an assembly of screens and diffusers, as in Fig. 6-25. These frequently get clogged from tiny particles or mineral deposits. You can simply take them apart and clean them out to remedy the flow problem. If needed, new aerators are readily available in most discount houses and building supply centers.

If the sprayer on your kitchen sink is plagued with low-flow problems, there are two probable causes. The first is a clogged screen or filter. Your unit probably has a screen or filter at the base of the handle. The small holes around the edge of the sprayer head may also be clogged.

The second probable cause may be the diverter valve in the faucet assembly. This is much more rare and is also more difficult to fix. You must remove the cover from the faucet assembly, which may involve turning off the water supplies, removing both faucets, the spout, and the cover. If you have a diverter valve, it will be in the center at the base of the spout. You can remove this and replace the seals (usually O-rings) and reassemble the unit.

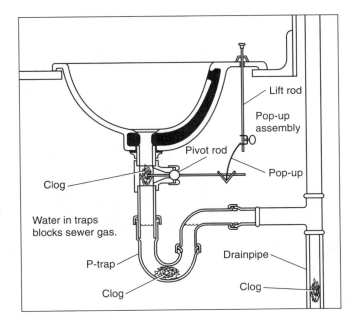

Fig. 6-26 *Common clog locations.*

Drains may be completely clogged and shut off all drainage, or they can be partially clogged and allow some drainage. Drains that are completely clogged are said to be *stopped* or *stopped up*. Drains that are partially clogged are said to be *slow* because the wastewater drains away very slowly.

If you understand how clogs are formed, then you can usually deal with them more efficiently. Grease is one of the most common causes of P trap clogs in kitchens. It is melted off dishes by hot washwater, but chills and congeals in the trap. The dark, moist environment is conducive to bacterial growth, and the grease provides the food for the bacteria. The result can be a dark, spongy growth that completely shuts off the drain. You can prevent a lot of this by simply running hot water down the drain for about 2 or 3 minutes each week.

Another cause of P trap clogs is sediment. If the sink is commonly used to wash dirt and mud off objects, the heavy particles of dirt will stay in the bottom of the trap. If you have a hobby, such as pottery, a P trap clog from the clay will be a common occurrence. The best way to deal with these clogs is to install a trap with a drain plug. Then you can periodically undo the plug and drain out the sediment.

Laundry drains, especially those for clothes washers, are another source of clogs caused by slow buildup of solids. The solids are tiny fiber bits from the fiber of the clothing being washed. They are combined with the detergent, which can act as grease or glue does to meld the fibers together. They will usually pass through the P trap, but can build up at a more distant turn. See Fig. 6-27. Preventive maintenance for a laundry drain is to

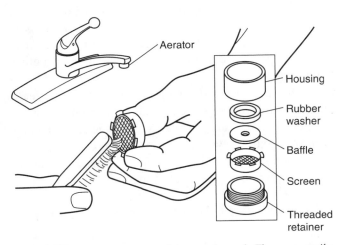

Fig. 6-25 *Faucet aerators often get clogged. They are easily cleaned or replaced.*

CLEARING DRAIN CLOGS

Clogged drains are one of the most common problems in a home. One reason is that there are lots of drains. You have them for toilets, sinks, lavatories, and laundries. The clog can be in the P trap, as in Fig. 6-26, or in the various drain lines inside or out of the house.

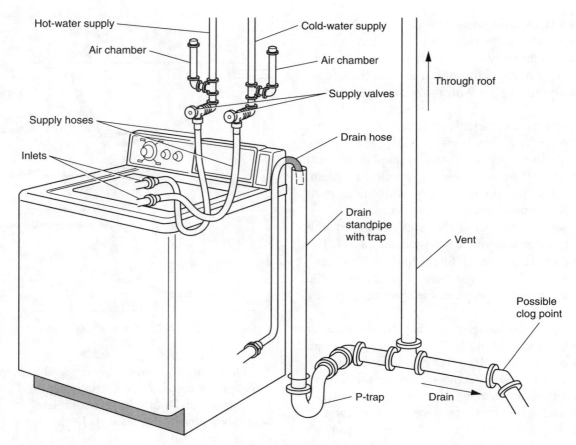

Hot-water supply

Air chamber

Cold-water supply

Air chamber

Through roof

Supply valves

Supply hoses

Drain hose

Inlets

Drain
standpipe
with trap

Vent

Possible
clog point

P-trap

Drain

Fig. 6-27 *Laundry clogs are often past the P trap.*

simply run an empty load or two of hot water, with no detergent, through the drain about once a month.

Then there are other clogs. Toilets, main drains, and even vents clog up. Poor drain design can also be a cause of clogs.

What to do You know you have a clog when the water drains away very slowly or when it doesn't drain at all. When you have a clog, there are several steps you can take. As a rule, you should start with the easiest first and progress toward the more difficult when the clog persists. A good sequence is to first consider any of the various chemical drain cleaners. If that doesn't work, then try a plunger. If a plunger can't do the job, then either a blow bag or an auger can be tried. If this doesn't work, then you need to either start taking stuff apart or call a plumber.

Chemical Drain Cleaners

Chemical drain cleaners, as in Fig. 6-28, are common household products. They are relatively inexpensive, easy to use, and found in a variety of stores from supermarkets to building supply centers. Many brand names

have a regular strength and an industrial-strength product. The main difference in the two strengths is the higher ratio of active ingredients in the industrial strength products. They are also available in regular and giant sizes.

Most are now made in liquid form, which is easier to use than the old two-stage crystals spooned into a drain. Either way, most are poisonous and toxic and can damage your fixture if you don't use them as directed.

Chemical drain cleaners can be used for just about every type of drain except toilets.

Clearing Clogs with a Plunger

The ubiquitous plunger, also called the "plumber's friend," should be a part of your tool array. If you don't have one, it is a wise investment. You can use a plunger on clogs in any type of sink, toilets, and many floor drains. It isn't effective on clothes washer drains.

There are three kinds with several variations of each. The most versatile type is the combination plunger, which features a bottom cone that can be used on toilets or folded up so the plunger can be used as a flat sink plunger. Figure 6-29 shows how it can be used

CLOGS
Maximum Strength for Your Worst Clogs

What Causes Clogs?

Most drain clogs are caused by the gradual accumulation of organic matter, hair, and grease on the inside walls of drainpipes. Over time, this buildup slows down the normal flow of water through the pipe. When a clump of debris or wad of hair tries to pass through the pipe, it can become stuck and completely stop the flow.

How Should I Treat Clogs?

If water backs up quickly each time the drain is used the problem is most likely close to the drain opening in the tub or sink.

Hair clogs often form around the stopper mechanism in the first vertical section (A).

Most clogs that form in the trap (B) are caused by foreign objects. While a drain opener can provide temporary relief by removing debris around the foreign object, the object will eventually have to be removed.

Clogs can also be caused by buildup in the lateral section of the pipe (C). Use a clog remover to treat a stopped or slow-running drain.

If the drain doesn't begin to back up immediately, the problem is farther in the drain system. Use a larger quantity of product (32 to 64 oz) to enable it to reach the clog site. If this is still not successful, a mechanical treatment (snaking or rodding by a plumber) may be required.

Fig. 6-28 *Try a chemical treatment first.*

Funnel cone at bottom

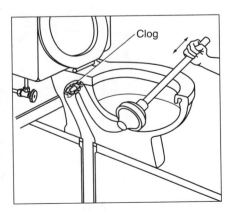

Clog

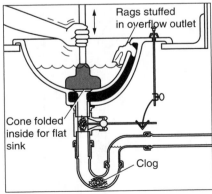

Rags stuffed in overflow outlet

Cone folded inside for flat sink

Clog

Fig. 6-29 *A combination plunger can be used to clear sinks or toilets.*

on sinks and toilets. It can also be used on tubs. The one difference is that the overflow hole on sinks or tubs should be plugged up with rags or plastic.

The way the plunger is used is the same for sinks, tubs, and toilets. The edge of the plunger is held firmly around the drain, and a sharp push is given to the handle. The downward thrust creates a surge of either air or water that is exerted against the clog. Then you lift the edge of the plunger to clear the vacuum and check to see if the fixture is draining.

If the clog persists, it is okay for the plunger cavity to fill with water. Simply hold the edge of the plunger firmly around the drain and repeat the process. A stubborn clog may take several thrusts. If you can't see any results after a dozen or so efforts, it's time to try something else.

Blow Bags

A *blow bag* is an effective tool for clogs in drainpipes, clothes washer drains, and some sinks. You attach it to a water hose and connect the hose to a hose bib faucet (either hot or cold). You can also buy adapters so that a hose can be attached to the spout of a kitchen sink.

The blow bag is inserted into the pipe, as in Fig. 6-30, and the water is turned on. The blow bag has a collapsible valve inside it that initially blocks the flow of water, causing the sides of the blow bag to expand. When the expansion firmly seals the bag against the walls of the pipe, the valve opens and shoots a surge of water against the clog. The action of the valve will create a series of powerful surges that work against a clog.

There are different sizes of blow bags. Each one will have only an inch or so variance in the size of the

pipe it can be used on. A typical size is one that can be used on 1½- to 2-inch pipe. One or two of these are also good additions to your tool collection. They are relatively inexpensive and are good as long as the sides remain flexible and without cracks.

Augers

Augers are another vital tool group for clearing clogs. They are often called *snakes*. As you might expect, there are a number of different types and sizes. It is a good idea to have at least one in your collection of tools. There are hand augers, power augers, and closet (or toilet) augers. Hand augers are recommended for the homeowner, while power augers are larger and more rigid augers are best suited for clearing main drain lines from the house to the sewer main.

Hand augers The handiest form of a hand auger is shown in Fig. 6-31. An auger with a ¼-inch diameter and a length of 25 feet will be sufficient for most home uses. The thin diameter will allow the auger to snake around sharp turns in traps, and the length will allow it to reach most clogs. The bore of the auger is pulled out of the housing by hand and pushed down the drain until you meet an obstruction (Fig. 6-31A). The auger is then turned, using the handle of the housing to clear the clog (Fig. 6-31B). These augers can be purchased at most discount houses and building supply centers.

Closet augers Closet augers are rather short and are designed to be used only on water closets (toilets). A ceramic toilet is easily stained when anything made of iron or steel is rubbed against it. To use a closet auger, you pull the auger back up against the plastic base, as in Fig. 6-32. Then you push the plastic base down to

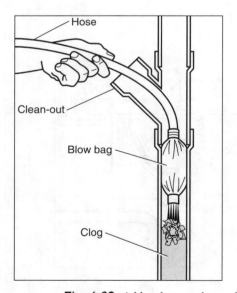

Fig. 6-30 *A blow bag can be used on clogs.*

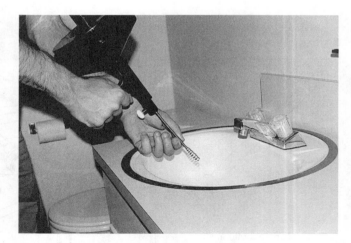

Fig. 6-31A *(A) The tip of the hand auger is pulled from the housing and inserted in the drain. The pop-up valve must be removed first.*

the base of the toilet trap as shown (Fig. 6-32B). This holds the auger clear of the ceramic until it touches a nonvisible area. The auger is then pushed through the handle to clear the clog.

Closet augers are medium-priced tools, for the most part. The purchase price of one is far less than

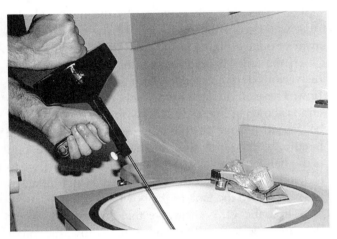

Fig. 6-31B *Push the auger as far as you can, then rotate the housing to spin the auger and clear the drain.*

one visit by a professional plumber, however. While a closet auger is probably not on the list of the first tools a homeowner should buy, it is nevertheless a good buy in the long run.

Power augers As previously mentioned, power augers are large, powerful tools suitable for clearing major clogs in main drains. They are a mainstay of professional plumbers. A homeowner who wishes to tackle this type of job should probably rent one rather than buy it.

If you tackle a main drain clog, first make sure that's where the problem is. You should have a clean-out plug or a house trap someplace. Loosen the cap slightly. If you don't have any waste seeping out, then loosen it some more. If you can remove the cap completely, then use a bucket or hose to run water into the drain. If it backs up, then your problem is in the drain. If it doesn't, then the clog is in the house.

If there is a clog in the drain, you can insert the auger into the pipe from the clean-out plug or the house trap. If the clog is beyond the reach of the auger, you can make an opening in the drain, as shown in Fig. 6-33.

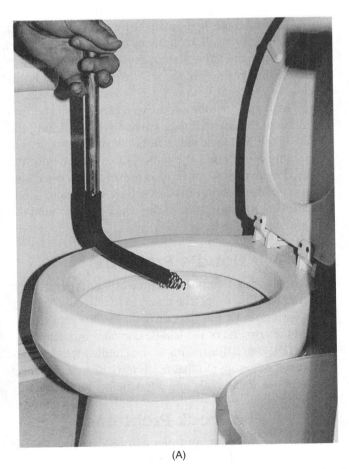

(A)

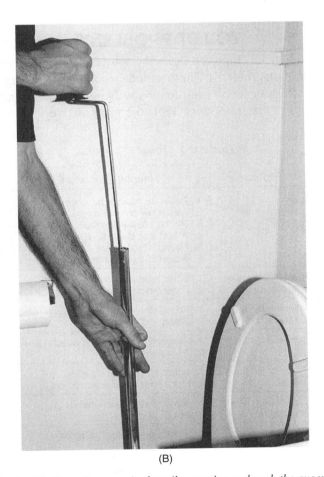

(B)

Fig. 6-32 *(A) Draw the end of the closet auger up against the plastic tip. (B) Place the auger in the toilet opening and push the auger through the tube as far as you can. Then turn the handle to clear the clog.*

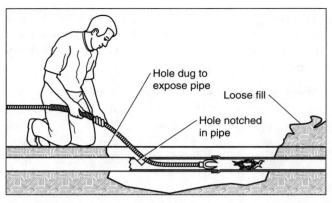

Fig. 6-33 *Tapping a drain or sewer line to clear a clog. The hole can be covered with plastic, felt, or metal and reburied.*

Then you can insert the auger into the new opening in the top of the pipe and proceed.

To close the opening once the clog is cleared, you patch the hole. To do this, simply lay a piece of rubber or gasket material over the hole and then cover it with a piece of galvanized sheet metal (like roof flashing). You can also clamp the pieces in place, but the weight of the earth when you refill the hole will usually hold the patch in place.

TOILET PROBLEMS

Aside from clogs, a toilet is the source of several other common problems. These include incomplete flushing, not flushing, noise from continuously running water, and leaks of water or odors. Table 6-1 gives an overview of

Table 6-1 *Toilet Problems*

Problem	Repair Procedure
Toilet runs continuously	1. Repair or replace ballcock 2. Adjust water level in tank 3. Replace leaky float ball 4. Adjust and clean flush valve 5. Replace flush valve 6. Adjust lift wires or lift chain
Toilet overflows or flushes sluggishly	1. Clear toilet clog 2. Clear main waste and vent stack clog
Toilet handle sticks or is hard to push	1. Adjust lift wires. 2. Clean and adjust handle.
Handle is loose.	1. Reattach lift chain or lift wires to lever. 2. Adjust handle.
Toilet will not flush at all.	1. Adjust lift chain or lift wires. 2. Make sure water is turned on.
Toilet does not flush completely.	1. Adjust water level in tank. 2. Adjust lift chain.
Water on floor around toilet	1. Tighten tank bolts and water connections. 2. Replace wax ring. 3. Replace cracked tank or bowl.

both problems and typical remedies. Toilets work by releasing a large amount of water, stored in the tank, into the bowl. The water level in the bowl rises and flows out the trap, creating a siphon for the bowl. This pulls the remaining water and wastes out through the trap. See Fig. 6-34. When the tank empties, the surge of water is ended, and the air in the drain will break the siphon. The ballcock valve refills the tank and also runs water from the inside rim of the bowl to wash away residue clinging to the sides of the bowl.

If the toilet is not clogged, the problems relate to the various devices used to cause and control the water flow. These are located in the tank. A typical tank system is shown in Fig. 6-35. The problems will typically be caused by one of three elements. The first is the handle, which starts the flushing action. The second is the ballcock valve, which controls the water flow to the tank. The third is the flush valve, which releases the water from the tank to the bowl.

The ballcock is the valve that controls the water flow into the tank. When the water is low, the valve allows water to flow into the tank. When the water reaches a preset level, the valve turns off the water.

The flush valve is opened when the handle is turned. This valve is held open by an air cavity that causes the valve to float and stay open until the tank is empty. Once the valve flap or ball returns to the valve seat, the tank refills and the pressure of the water holds the valve in place.

The overflow tube serves two functions. The first is to keep a faulty ballcock valve from overfilling the tank. If the ballcock valve fails (as many do), it will run continuously and the tank will overflow onto the floor. The overflow valve also prevents this overflow. The second function of the overflow valve is to guide water from the refill tube to the water ports around the inside edge of the bowl.

Handle Problems

One of the most common sources of problems is the handle. See Fig. 6-36. The chain to the flush valve may be sticking under the flapper (or ball). Sometimes the eye at the arm end of the handle corrodes away, and the chain falls out. Adjustments to the handle include tightening the retaining nut that holds it to the tank, bending the arm inside the tank, and adjustments to the chain.

Ballcock Problems

A continuously running toilet can be caused by flush valve problems, but is more likely caused by incorrect water level settings or by ballcock problems. There are

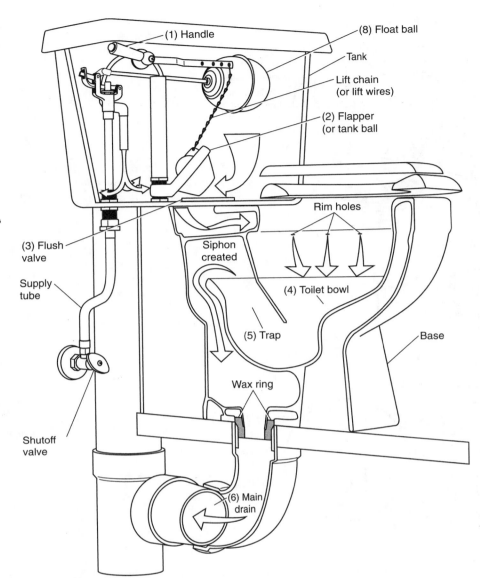

(1) Handle

(8) Float ball

Tank

Lift chain
(or lift wires)

(2) Flapper
(or tank ball)

(3) Flush
valve

Supply
tube

Rim holes

Siphon
created

(4) Toilet bowl

(5) Trap

Base

Wax ring

Shutoff
valve

(6) Main
drain

Fig. 6-34 *Toilet action: When the handle (1) is pushed, the lift chain raises a flapper (2), water rushes through the flush valve (3) into the toilet bowl (4), and waste water in the bowl is siphoned from the trap (5) into the main drain (6). When the tank is empty, the flapper seals the valve, and the ballcock (7) refills the toilet tank. The ballcock is controlled by a float (8) that rides on the surface of the water. When the tank is full, the float ball shuts off the ballcock.*

four types of ballcocks commonly used: plunger valve, cartridge (floatless), diaphragm, and float cup.

Plunger valve ballcock The oldest type of ballcock used is a solid brass unit actuated by a float ball on the end of a rod. See Fig. 6-37. This type of system can usually be repaired. The unit is a variation of a globe valve and has washers that can be changed around the stem and next to the seat. You would fix it in much the same as any globe-type valve. Turn off the water supply, take apart the top part of the unit, replace the washers, and so on.

About the only other adjustment is to bend the arm holding the float ball to adjust the water level. Most toilet tanks have a line at the back that marks the best water level for that unit. If the water level is set too high, water will spill into the overflow tube and the toilet

will run continuously. If the level is set too low, the toilet will not flush completely.

Cartridge (floatless) ballcock This unit is a small plastic cartridge near the bottom of the tank. In operation, it is completely immersed in water. This unit can be adjusted for different water level settings by an adjustment screw, as shown in Fig. 6-38. If adjusting the water level doesn't fix your problem, you usually need to replace the ballcock. Further, most codes no longer allow this type of ballcock, so you must replace it with an acceptable type.

Diaphragm ballcock This type of ballcock, shown in Fig. 6-39, was once made from brass. Many of these are still around. Most of the newer ones are now made of plastic. Technically, both types are repairable, but finding parts is a problem. Only a major plumbing supply

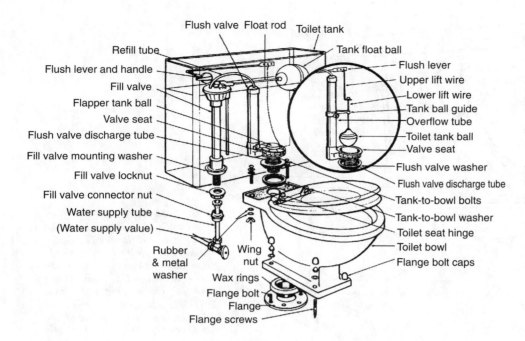

Flush valve Float rod Toilet tank
Refill tube
Flush lever and handle
Fill valve
Flapper tank ball
Valve seat
Flush valve discharge tube
Fill valve mounting washer
Fill valve locknut
Fill valve connector nut
Water supply tube
(Water supply value)
Rubber & metal washer
Wing nut
Wax rings
Flange bolt
Flange
Flange screws

Tank float ball
Flush lever
Upper lift wire
Lower lift wire
Tank ball guide
Overflow tube
Toilet tank ball
Valve seat
Flush valve washer
Flush valve discharge tube
Tank-to-bowl bolts
Tank-to-bowl washer
Toilet seat hinge
Toilet bowl
Flange bolt caps

Fig. 6-35 *The tank system controls the water flow.*

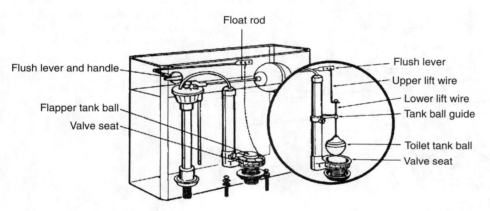

Float rod
Flush lever and handle
Flapper tank ball
Valve seat

Flush lever
Upper lift wire
Lower lift wire
Tank ball guide
Toilet tank ball
Valve seat

Fig. 6-36 *Handles cause many problems.*

or building center will have these parts, if at all. The typical hardware or department store will not likely carry parts for this type of ballcock. The plastic ones are only slightly more expensive than the float cup ballcock, and you normally just replace them.

Float cup ballcock   This type, shown in Fig. 6-40, is becoming very common. It is inexpensive, reliable, and easy to replace. The float level (and water level in the tank) can be adjusted by moving the spring clip on the pull rod. When this type of unit fails, water will run continuously. Water may spill from the sides or simply run from the ports at the bottom of the unit.

Technically, these units can be repaired, but parts are usually not available in most stores. In reality, when this type of unit fails, you must replace it.

Ballcock replacement   To replace any of the ballcock types, the same basic procedure is used. First, turn off the water supply at the cutoff valve. Flush

the toilet and hold the handle down until no more water runs. Then remove the top of the tank and put it in a secure location. It is fragile. Sponge out the water remaining in the tank. Next, disconnect the supply line, as in Fig. 6-41, by unscrewing the supply coupling from the stem of the ballcock. Then unscrew the mounting nut that holds the ballcock in place. Remove the refill tube and pull up the old ballcock.

To install the new valve, place a new cone washer on the valve stem. The cone should taper down. Secure it in place with the mounting nut and reattach the supply line. Next, attach the refill tube to the overflow tube. If you forget and leave it loose, water will splash all over.

Once everything is secure, turn the water back on and check everything for leaks. Tighten any connections that leak. Then adjust the water level and flush the toilet two or three times to check everything. Once the system is at the correct water level and there are no leaks, replace the tank top.

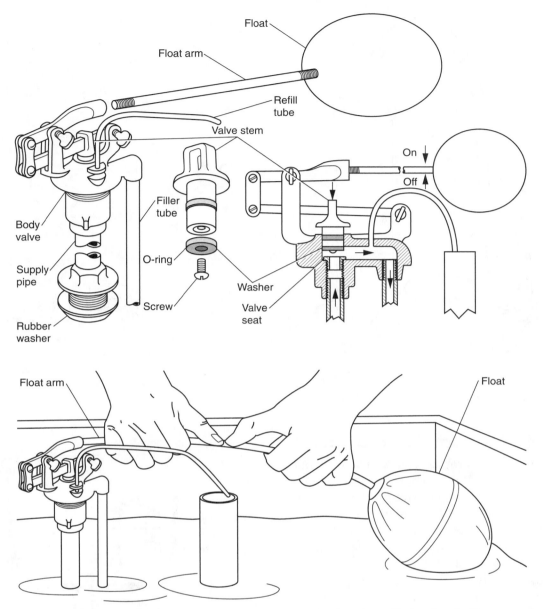

Fig. 6-37 *Traditional plunger valve ballcock. Tank water level is adjusted by bending the float arm.*

Flush Valve Problems

The most common causes of flush valve problems stem from the deterioration of the float unit that is the valve seal. You can identify this problem from an intermittent opening of the ballcock. You will probably also hear a very quiet trickle. The valve may be a ball or a flap, as in Fig. 6-42. To check it, turn the water off and flush the toilet. Then check the valve for rough edges or cracks. If you find these, simply replace the valve. This will usually solve the problem.

Sometimes the action of the flowing water will cause the lip of the valve seat to deteriorate. You can usually identify this by running your finger around the edge. If you feel a groove or a rough spot, particularly if it is discolored, you may need to reface the valve seat. You do this while the tank is empty.

Once the valve seat is exposed, remove the ball or flapper. Find a flat block of wood that covers the entire opening and place a wide sheet of fine-grit sandpaper across the full face of the block. Place the block and sandpaper across the face of the seat as in Fig. 6-43, and move it back and forth with a very small motion. Do this until the roughness, discoloration, or groove is gone. If the problem persists, then you will need to take apart the tank and replace the overflow tube/flush valve assembly. This procedure is covered a bit later in this chapter.

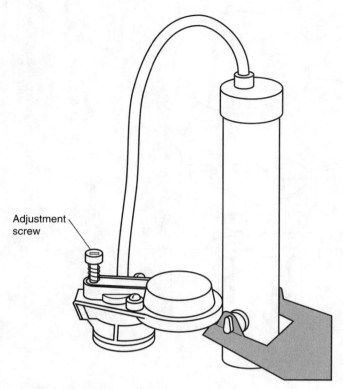

Fig. 6-38 *A cartridge ballcock has no float.*

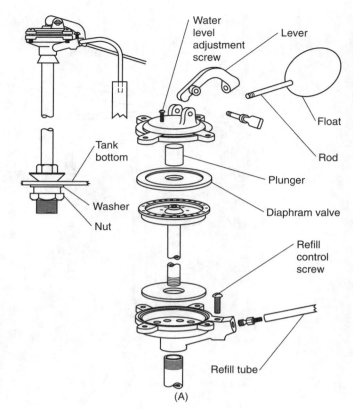

(A)

Toilet Disassembly Problems

There are a number of problems that can be fixed only after the toilet is partially or completely disassembled. Leaks around the base of the tank can be caused by gasket or washer failures. There are three locations for this. See Fig. 6-44. The first is under the ballcock valve. You can usually fix that one by just tightening the mounting nut under the tank.

The other two types of leak often require some disassembly of the tank. The first type is leaks from the mounting bolts that connect the tank to the seat. The other leak can come from the spud washer or cone washer that seals the flush valve and bowl.

Tank washers   Sometimes you can tighten these and stop the leak, but bad washers cause a large portion of these leaks. Some toilets have two mounting bolts, and others have four. Feel beneath the tank to locate drips or wet spots caused by these washers. Next, check the inside of the tank to see if the washers are deformed.

Try tightening the nuts under the flange on the bowl just a little. You don't want to get too much pressure on these, or you will break the mounting flange on the bowl. Remember it is a ceramic product just about like a coffee mug. If slight tightening doesn't work, it's a good idea to replace the washers. Washers are typically bad if they appear rough or cracked or if they give off clouds of color when touched.

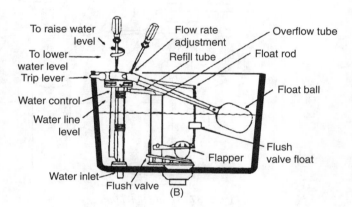

Fig. 6-39 *Diaphragm ballcock.*

Empty the tank just as you would to change a ballcock. Unscrew the nuts under the bowl flange, as in Fig. 6-45, and remove the bolts, metal washers, and soft washers. These may be any color rubber or neoprene. You can buy kits that have new bolts and washers, or you can buy just the washers. It's your choice.

Replace the tank washers and place the bolts back in the tank. It's a good idea to place a soft washer between the metal washer and the toilet bowl flange. Then replace the nut and tighten it until the tank washer begins to compress. Turn the water back on and refill the tank. Check for leaks as soon as water covers the bottom of the tank. Tighten the bolts as little as you can, but tighten them until there are no leaks.

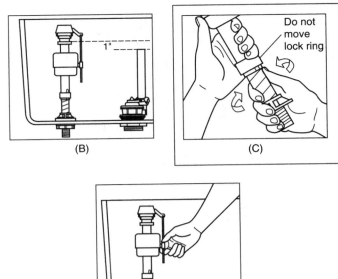

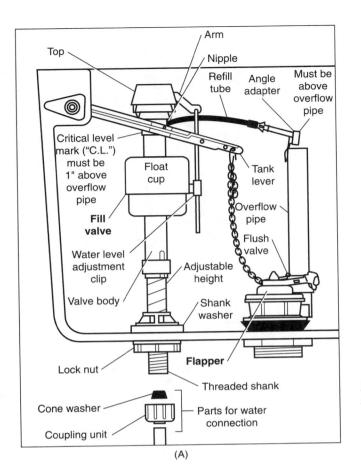

Fig. 6-40 *(A) Float cup ballcock is now very common. (B) and (C) Its height is adjusted for the tank. (D) The clip adjusts the water level.*

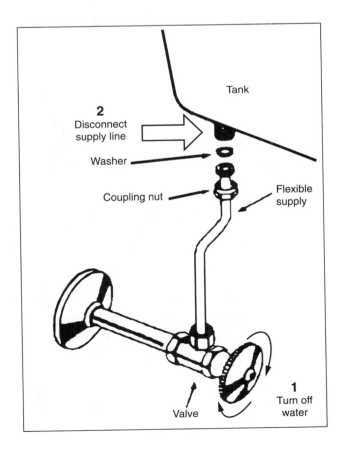

Fig. 6-41 *To replace a ballcock, turn off the supply valve and disconnect the supply line at the tank.*

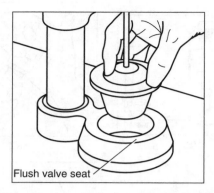

Flush valve seat

Fig. 6-42 *Feel the lip of the flush ball or flapper. It should be soft, smooth, and even.*

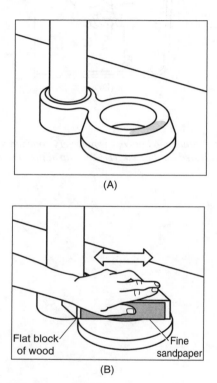

(A)

Flat block of wood   Fine sandpaper

(B)

Fig. 6-43 *Resurface a flush valve seat with a sanding block.*

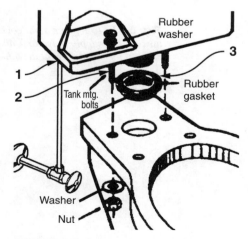

Rubber washer

1

2

Tank mtg. bolts

3

Rubber gasket

Washer

Nut

Fig. 6-44 *Tank leaks have three sources.*

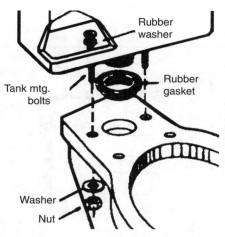

Rubber washer

Tank mtg. bolts

Rubber gasket

Washer

Nut

Fig. 6-45 *Tank leaks can be from around the mounting bolts or the gasket. Try tightening the bolts.*

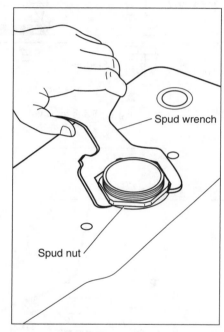

Spud wrench

Spud nut

Fig. 6-46 *Use a spud wrench to remove the spud nut on the bottom of the tank.*

Flush valve washer Sometimes these washers go bad. Like any other washer, they crack and peel with age. They also become dry, brittle, and deformed. When this type of washer goes bad, you will notice a leak that comes from under the center of the tank.

To replace it, you must turn off the water, empty the tank, and remove it. Remember, it is fragile.

Use a spud wrench to remove the spud nut holding the flush valve/overflow tube to the tank. You may find one or more washers, as shown in Fig. 6-46. When you remove the tank, you may find a spud washer or a cone washer, as in Fig. 6-47.

The best and surest way to fix this type of leak is to replace the overflow tube/flush valve seat assembly

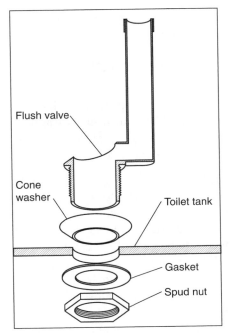

Fig. 6-47 *The flush valve attaches to the tank with a cone washer, gasket, and spud nut.*

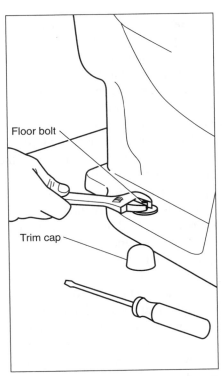

Fig. 6-48 *Begin to remove the stool by removing the trim caps and unscrewing the nuts on the floor bolts.*

and all the washers. You can also use bathroom caulk between the flush valve and the inside of the tank. Because you had to take the tank off, it's a good idea to replace the tank washers as well.

Wax Ring Problems

Toilets are sealed with a wax ring between the stool and the floor. Refer to Fig. 6-34. These turn bad from a variety of causes. The wax can melt from heat, shrink with age, or be deformed by a toilet poorly leveled or secured.

You detect a wax ring problem when you find wastewater seeping from underneath the toilet bowl, or when there is a bad odor in the room. Either or both conditions may be found. When the wax is separated from the bowl, water may not leak out. However, because the toilet trap is built into the toilet bowl, there is no seal from the main sewer or drain line odors. Thus, you may detect a foul odor before you notice anything else.

To replace a wax ring, you shut off the water, empty the tank, and remove it from the stool. You also remove the toilet seat. Then you undo the floor bolts at the base of the stool, as in Fig. 6-48. Next, you lift up the stool and turn it upside down. It's a good idea to stuff the opening in the floor with rags, paper, or plastic to keep the odor out.

The wax ring may be on the floor or stuck to the bottom of the toilet. Either way, scrape the ring residue off both surfaces with a putty knife, or something similar. Then clean up any traces of sealant around the base of the stool.

If you are replacing a toilet, there are several factors to consider.

Toilet Selection

Toilets, or water closets, come in several different mechanisms and styles. The styles include those that fit in corners (Fig. 6-49), rest on the floor, and have their weight supported entirely by a wall. Corner toilets are designed to save space and are particularly functional

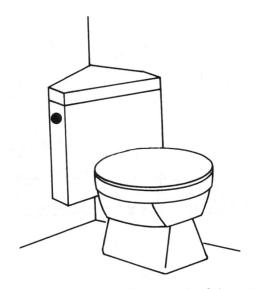

Fig. 6-49 *A space-saving toilet designed to fit into a corner.*

in very small rooms. Even the triangular tank fits into a corner to save space. The wall-hung toilet is expensive and requires sturdy mounts in the wall. The most common is the floor-mounted toilet (Fig. 6-50).

Floor-mounted toilets can be obtained in different heights. Older people generally find that an 18-inch-high toilet is easier to use than the 14-inch height of conventional units. Heights range from 12 to 20 inches, and some can be purchased with handles and other accessories for the ill or handicapped.

The siphon jet is perhaps the most common type and is most recommended. It is quiet and efficient. Most of the bowl area is covered by water, making it easier to clean. The siphon action mechanism is an improvement over the siphon jet. It leaves no dry surface, thus making it easier to clean. It is efficient, attractive, and almost silent. It is also the most expensive. Most builders recommend the siphon jet because it costs less.

Up-flush toilets are used in basements when the main sewer line is above the level of the basement floor. These require special plumbing and must be carefully installed.

Generally, the better the quality, the higher the cost. Assuming three grades, the cheapest will not be made to withstand long, heavy use. The difference between medium and high quality will be the thickness of the plating and the quality of the exterior finish.

Toilets, bidets, and some lavatories are made of vitreous china, which is a ceramic material that has been molded, fired, and glazed, much as a dinner plate has. This material is hard, waterproof, and easy to clean and resists stains. It is very long-lasting; in fact, some china fixtures are still working well after 100 years or more. White is the traditional color, but most manufacturers now provide up to 16 additional colors.

Shopping around can give many insights into colors and features available.

Toilet Installation

Installing the two-piece toilet requires some special attention to details, to prevent leakage and ensure proper operation. The unit itself is fragile and should be handled with care to prevent cracking or breaking. Keep in mind that local codes have to be followed.

Roughing in Use Fig. 6-50 as a reference. Notice the distance from the wall to the closet flange centerline. The distance varies according to the unit selected. For instance, American Standard's 4010 tanks require the distance A to be 25 millimeters or 10 inches. Model 4014 needs 356 millimeters or 14 inches for distance A. All other tanks require 12 inches or 305 millimeters. The tank should not rest against the wall. Also notice the location of the water supply.

Install the closet bolts as shown in Fig. 6-51. Install the closet bolts in the flange channel, and turn 90° and slide into place 6 inches apart and parallel to the wall.

Distance A shown here is the same as that in Fig. 6-50. Next, install the wax seal (see Fig. 6-52). Invert the toilet on the floor (cushion to prevent damage). Install the wax ring evenly around the waste flange (horn), with the tapered end of the ring facing the toilet. Apply a thin bead of sealant around the base flange.

Position the toilet on the flange, as shown in Fig. 6-53. Unplug the floor waste opening, and install the toilet on the closet flange so the bolts project through the mounting holes. Loosely install the retainer washers and nuts. The side of washers marked "This side up" *must* face up!

Install the toilet, per Fig. 6-54. Position the toilet squarely to the wall; with a rocking motion, press the

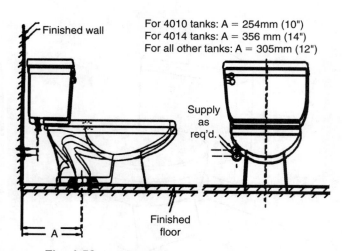

Fig. 6-50 *Roughing-in dimensions.* (American Standard)

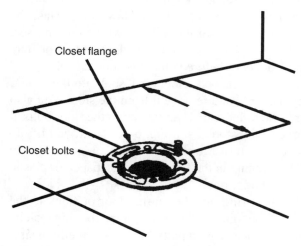

Fig. 6-51 *Closet flange and bolts.* (American Standard)

Fig. 6-52 *Installing the wax seal.* (American Standard)

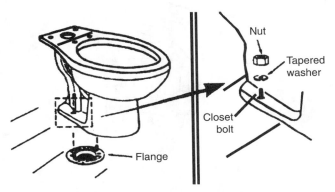

Fig. 6-53 *Positioning the toilet on the flange.* (American Standard)

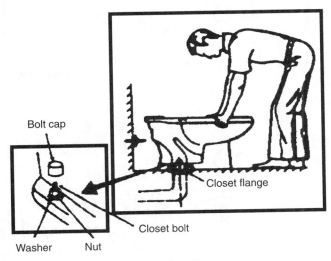

Fig. 6-54 *Installing the toilet.* (American Standard)

and if necessary, cut the bolt height to size before installing the caps. Smooth off the bead of sealant around the base. Remove any excess sealant. Next, install the tank. In some cases, where the tanks and bowls use the Speed Connect System, the tank mounting bolts are preinstalled. Install the large rubber gasket over the threaded outlet on the bottom of the tank, and lower the tank onto the bowl so that the tapered end of the gasket fits evenly into the bowl water inlet opening (see Fig. 6-55) and the tank mounting bolts go through the mounting holes. Secure with metal washers and nuts. With the tank parallel to the wall, alternately tighten the nuts until the tank is pulled down evenly against the bowl surface. *Caution!* Do not overtighten the nuts more than required for a snug fit.

In those instances where the bolts are not preinstalled, start by installing large rubber gaskets over the threaded outlet on the bottom of the tank. Then lower the tank onto the bowl so that the tapered end of the gasket fits evenly into the bowl water inlet opening. See Fig. 6-56. Insert the tank mounting bolts and rubber washers from the inside of the tank, through the mounting holes, secure with metal washers and nuts. With the tank parallel to the wall, you can alternately tighten the nuts until the tank is pulled down evenly against the bowl surface. Again, caution is needed to make sure the nuts are not overtightened. Install the toilet seat according to the manufacturer's directions.

Connect the water supply line between the shutoff valve and tank water inlet fitting. See Fig. 6-41. Tighten the coupling nuts securely. Check that the refill tube is inserted into the overflow tube. Turn on the supply valve and allow the tank to fill until the float rises to the shutoff position. Check for leakage at the fittings; tighten or correct as needed.

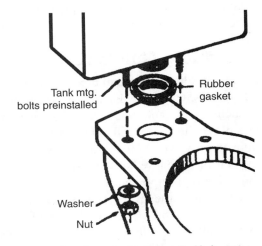

Fig. 6-55 *Installing the tank with preinstalled bolts.* (American Standard)

bowl down fully on the wax ring and flange. Alternately tighten the nuts until the toilet is firmly seated on the floor. *Caution!* Do not overtighten the nuts, or the base may be damaged. Install the caps on the washers,

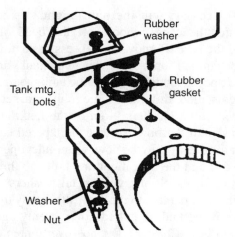

Fig. 6-56 *Installing the tank without preinstalled bolts.* (American Standard)

Adjustments   There are some adjustments that need to be made in most installations to ensure proper operation. Refer to Figs. 6-37 through 6-40.

1. Flush the tank and check to see that the tank fills and shuts off within 30 to 60 seconds. The tank water level should be set as specified by the mark on the inside of the tank's rear wall.

2. To adjust the water level, turn the water level adjustment screw counterclockwise to raise the level and clockwise to lower the level.

3. To adjust the flow rate (tank fill time), turn the flow rate adjustment screw clockwise to decrease the flow rate. This increases the fill time. Turn the adjustment screw counterclockwise to increase the flow rate or decrease the fill time.

4. Carefully position the tank cover on the tank.

5. The flush valve float has been factory-set and does not require adjustment. Repositioning the float will change the amount of water used, which might affect the toilet's performance.

Adding Cutoff Valves

If you are in an older home and are making upgrades as you do repairs, you may wish to add cutoff valves to fixtures such as sinks and toilets. It was quite common in past years to build homes without these conveniences. By adding cutoff valves, you eliminate the need to turn off the water for the entire house when you need to fix a running toilet or a drippy faucet.

First, look underneath the fixture and locate the supply line for the fixture. There will probably be a ½- or ¾-inch supply line jutting out a short distance. It may protrude from the floor or the wall. Next determine

the type of pipe, such as copper, galvanized steel, or plastic. Then determine what you need to do. Will you need to cut a copper or plastic line, or can you just unscrew a fitting from a galvanized pipe? Once you determine what you have, purchase the needed fittings. You can purchase a valve that will screw directly onto galvanized pipe. There is also a cutoff valve with a compression fitting that attaches directly to a copper line. For plastic pipe, a transition fitting that joins the pipe on one end and has a thread on the other end is needed.

If you must cut the line, as in Fig. 6-57, use a saw or tubing cutter. Then you attach the cutoff valve, as

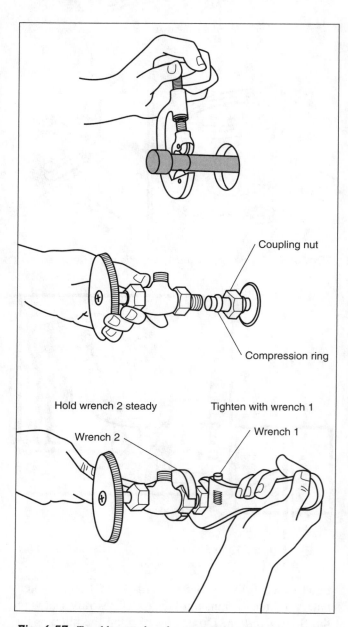

Fig. 6-57 *To add a supply valve to copper pipe, cut the old pipe. Then use a compression fitting as shown.*

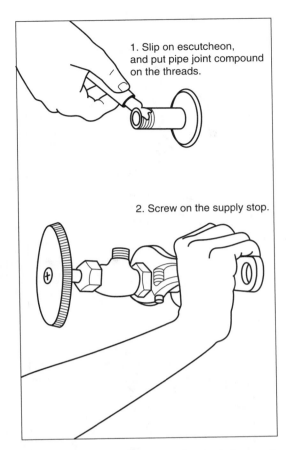

1. Slip on escutcheon, and put pipe joint compound on the threads.

2. Screw on the supply stop.

Fig. 6-58 *Screw a supply valve onto galvanized pipe or a threaded transition piece for plastic pipe.*

shown in Fig. 6-58, and attach the supply lines from the fixture. If you are replacing the supply line too, then you might consider the flexible hoses shown in Chap. 3.

REPAIRING PIPE DAMAGE

A homeowner occasionally must fix a bad pipe. Damage can be caused by corrosion from either the inside or the outside. Pushing or hitting a pipe can cause the joints to break or leak. Freezing weather can cause the water in a pipe to freeze and expand, to split a pipe or break a joint. On rare occasions, a bubble or fault within the metal or plastic of the pipe can cause the pipe to fail. Whatever the reason, when pipe failure occurs, water spurts out and repairs are required.

There are different ways of coping with something in the middle of the night, and there are techniques for large leaks and small leaks. First, let's consider the temporary methods that might be used in the middle of the night.

Temporary pipe repairs When a pipe fails in the middle of the night, there are several things to con-

sider. There are probably no stores open, so you must make do with what you have.

The first order of business, of course, is to turn off the water. Then assess the extent of the damage. If you have a drip from a very tiny opening, the easiest thing to do is to wrap the area with several layers of electrical tape. It's also a good idea to add a clamped cover, as shown in Fig. 6-59. You can make this from a couple of hose clamps and a piece of a soda can.

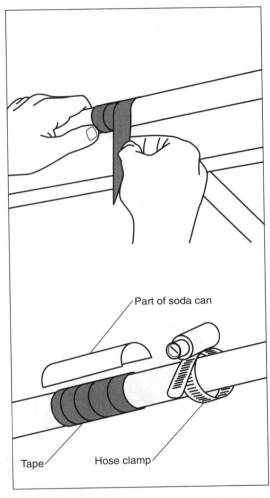

Part of soda can

Tape

Hose clamp

Fig. 6-59 *Wrap electrical tape over a tiny leak. Then clamp metal over it with a hose clamp.*

Small holes in metal pipe can be plugged with a sheet metal screw and a soft washer, as in Fig. 6-60. A soft washer can be made from a piece of rubber, gasket material, or a faucet washer. Always use a screw that won't penetrate the backside of the pipe. You can plug the hole with a wooden plug, such as a toothpick or a pencil point, as in Fig. 6-61. The plug may work, but using a clamp of some kind works better. You don't

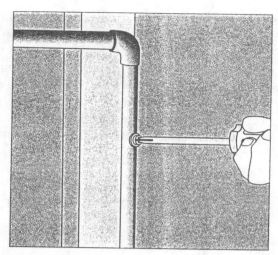

Fig. 6-60 *A small round hole can be plugged with a soft washer and a short screw.*

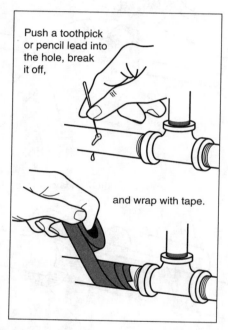

Push a toothpick or pencil lead into the hole, break it off,

and wrap with tape.

Fig. 6-61 *Plug a small hole with a toothpick. Then wrap it with tape and add a clamp, as in Fig. 6-59, for best results.*

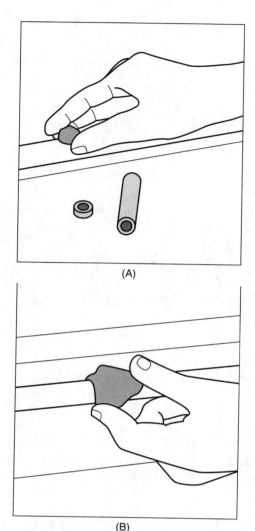

(A)

(B)

Fig. 6-62 *Plumber's epoxy putty works well. Cut off a small amount, knead it, and press it into the leak. Then add some all around the pipe.*

need to tighten the clamp very hard—just enough to hold the plug in place when you turn the water on. Another trick is to coat the plug with epoxy, wait about an hour, and then turn the water on. Again, it works better with a clamp.

Another quick way is to use epoxy putty (Fig. 6-62). There are two or three types of the epoxy, but the handiest is like a small roll of putty. The resin is in the core of the roll, and the hardener is on the outer surface. First clean the surface around the leak with an abrasive. Then simply break off a small amount of the putty,

knead it together to activate the hardener, and rub it into and over the leak. Allow it to harden before you turn the water back on. Then check it to be sure it's working.

Next, you can use a leak clamp. There are two versions, a factory-made one and a hastily rigged middle-of-the-night home version. Both are shown in Fig. 6-63. For a lengthy leak, a version of the rubber and tin can strategy can work. The key to making this work is to use several hose clamps along the length of the patch. See Fig. 6-64.

To end this section, some suggestions of what to keep on hand are appropriate. Epoxy putty will keep for months, if not years. It's highly recommended. Several extra hose clamps are also on the list. You should get diameters that are about 1 inch more than your pipe diameter. Remember that the outside diameter of your pipe is greater than the official size of the pipe. So, if

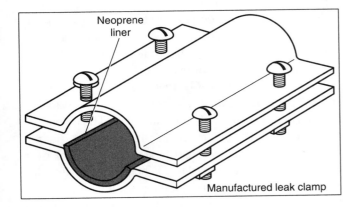

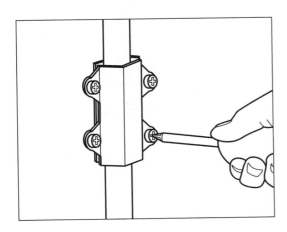

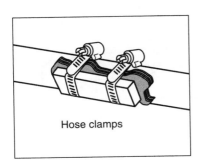

Hose clamps

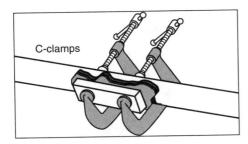

C-clamps

Fig. 6-63 *Temporary patches over large leaks can be done with leak clamps.*

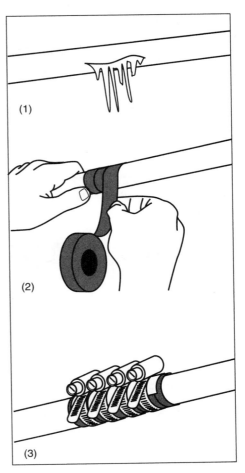

Fig. 6-64 *Fix a long leak with tape, tin, and hose clamps.*

you have ³/₄-inch galvanized supply lines which have an outside diameter of about 1 inch, you will want clamps good for up to 2 inches.

If you can get some rubber or neoprene scraps, they are nice to have. If you don't have any scraps, buy the smallest, cheapest bicycle tire you can. It won't cost much, and it makes excellent patch pieces. You can also use gasket material that you can get at an automotive supply house. If you can, buy a couple of the pipe leak clamps, shown in Fig. 6-63. If you can't get those, consider a couple of C clamps of sufficient capacity.

Thawing Pipe

Sometimes pipes freeze in places that you can reach. It's common for pipes to freeze around basement walls and in the outside walls of a house. You can also get a freeze if a space is open to outside air. In the initial stages, a pipe will usually freeze in one place first, with the freezing action spreading from there. At this stage, you can thaw the freeze without pipe damage.

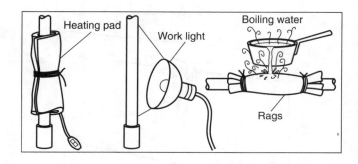

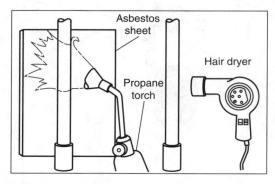

Fig. 6-65 *Ways of thawing a frozen pipe. Don't use a torch on plastic pipe.*

The main difference is that one is soldered and the other is cemented. To fix a pipe leak, cut an inch or so past the leak on each side. A tubing cutter works well if there is room. If not, a reciprocal saw or a mini hacksaw is a good tool. Don't try to cut near the damaged pipe. There may be more damage than you see, so cut a little extra out of the good pipe to be sure you get all the damage removed. Get rid of the burrs, and then measure the distance you need to replace, as in Fig. 6-66.

There are two options. The first is to use *slip couplings,* as in Fig. 6-66B. In some areas you can buy special couplings called slip couplings. These do not have a central ridge or shoulder in them. Thus, you can slip them well past the break onto the good pipe, put the new pipe section in place, and slide the fittings back over the joint and fix them in place. When you use slip couplings, it's a good idea to mark the place where you want to fix them in place. Once you put both slip couplings over the joints, you can't see exactly where they are in

Freezing pipes are a problem because water expands when it freezes. The freezing action can eventually result in pipe damage, as previously mentioned.

Anything you can do to put heat on the pipe will work. One warning, however. Don't use a torch or any type of high heat on plastic pipe. See Fig. 6-65. Hot water can be poured on the pipe. Wrapping the pipe in rags to hold the water will help retain the heat. Hair dryers and lamps will work. While a heat lamp is good, any electric lamp will work. Heating pads and strips are good solutions, too. You can even wrap the lamp next to the pipe with foil. The main idea is to get heat to the frozen area. Then wait. It may take several minutes.

Permanent Pipe Repairs

When you replace part of a pipe system because of pipe failure, the basic ideas are about the same regardless of the type of pipe. To effect a permanent repair, you must replace the faulty section. There are two basic ideas: (1) Fix a break in a pipe, and (2) fix a bad fitting or joint. To fix a pipe break, you must cut out the damaged portion and replace it. To fix a bad fitting, you usually have to cut at least one of the pipes leading to it. Actual practice varies with the type of pipe and accessibility. These two ideas will be presented for each pipe type.

Copper and plastic pipe Repairs for both copper and plastic pipe are done in almost the same manner.

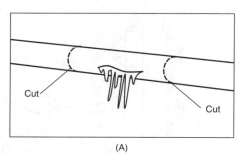

(A)

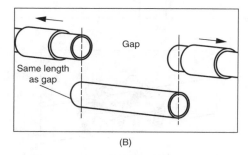

(B)

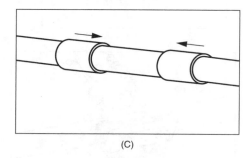

(C)

Fig. 6-66 *Fixing a leak with slip fittings. (A) Cut out the leaking section of copper or plastic. (B) Cut a new piece the same length as the gap. Slide slip couplings on each side. (C) Apply cement or solder and slide in place.*

respect to the ends of the pipe. A mark to align each fitting is a big help. Slip couplings are becoming scarce and are simply not available in some regions.

The second method is to use a *union* as in Fig. 6-67. Remember that you can buy unions for either copper or plastic pipe. Each type has a shouldered fitting on each end so the union can be soldered or cemented to the pipe. Again, you cut away the damaged pipe and a bit more. Measure the total distance you must replace. Then you allow for the width of the union and the shoulder depth of the fittings. Next, you cut nipples of the appropriate length and fix them in place. Take apart the union, slide the threaded nut onto one pipe, and attach the unthreaded part of the union to that pipe. Then attach the threaded part of the union to the other side of the pipe, as shown. Screw the two parts together to complete the repair.

You can use compression unions, as discussed in Chap. 3, but you will need two for a break of more than about 1/2 inch. Compression unions, however, are not regarded as a permanent repair in all parts of the country.

To fix a leak in a fitting, such as an elbow, you must remove the elbow. While you can sometimes just resolder a copper fitting, it usually doesn't happen. The reason is that often there is just enough water in it somewhere to keep you from getting enough heat on the fitting to fuse the solder. In the long run, it is generally quicker and easier to make the cuts shown in Fig. 6-68 and make the repairs as indicated.

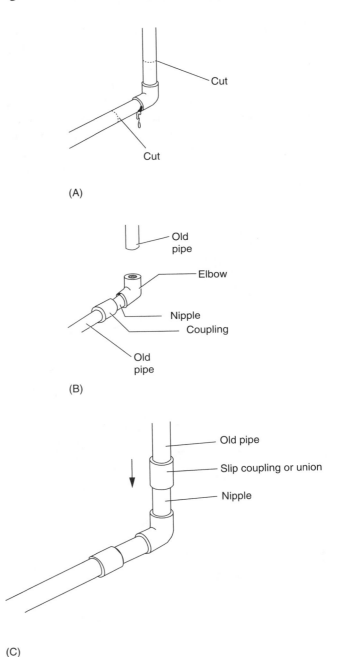

(A)

(B)

(C)

Fig. 6-68 *Fixing a leaky fitting for copper and plastic pipe. (A) Cut out the bad fitting. (B) Build up one side with standard fittings. (C) Use nipple and union or slip coupling to complete the job.*

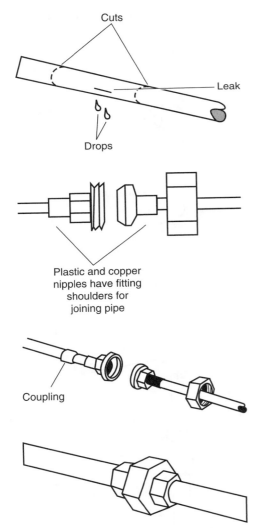

Fig. 6-67 *Unions are made for copper and plastic pipes.*

Repairing Galvanized Steel Pipe

The basic idea in fixing a leak or a fitting on galvanized pipe is about the same as for other pipe. However, the major difference is due to the way the pipe is assembled by threading. You cannot unscrew a piece of pipe from the middle. To unscrew the pipe, you must start at one end and disassemble back to your trouble point. Of course, this is simply not possible in most situations. The solution is to cut the pipe.

As a rule, you need to cut only one pipe. The rest of the problem area can be unscrewed. To fix a pipe leak, follow the procedure shown in Fig. 6-69. First, make a cut an inch or so away from the visible leak. The handiest tool for this will probably be a reciprocal saw. A pipe cutter will work if there is room, and a hacksaw or mini hacksaw is the next choice. Next, unscrew the damaged part. If it is not very far to a fitting on both sides of the leak, you can replace the damaged pipe with two nipples and a union. If the leak is in the middle of a long pipe, you should probably just cut it out.

You will use a nipple to replace the damaged area as shown and a union to join the cut sections, as shown in Fig. 6-69. However, you must thread the end of the pipe you left in place. If the length of the pipe to the next fitting is visible, you can unscrew the pipe and thread it on a workbench or other convenient place. If not, you can use a pipe die and die wrench to thread it in place. If access is limited, use the ratchet feature of the die wrench. Then assemble the repair as shown.

If you run into a stubborn fitting that just won't unscrew, loosen it first with a penetrating lubricant and tap it slightly. If that doesn't work, try heating it with a torch, as shown in Fig. 6-70. A minute or so of heat will cause the fitting to expand away from the pipe. Even just a little expansion will likely break loose the thread enough to unscrew. You should remember that the fitting is hot and touch it only with a wrench.

To repair a leaky fitting, usually an elbow, you must again cut one pipe. The best choice is to pick a short area between two fittings, as shown in Fig. 6-71.

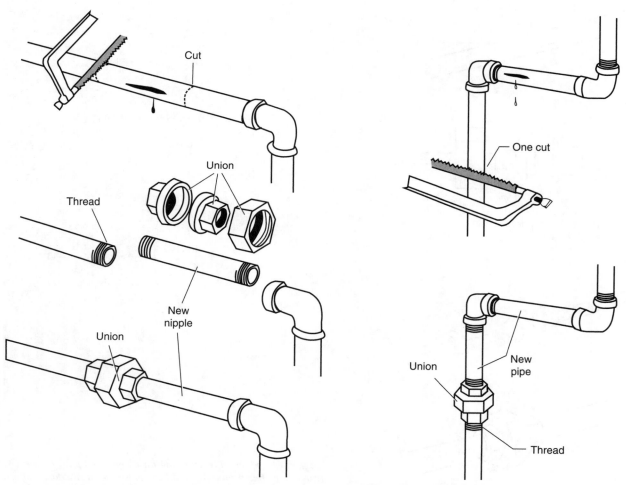

Fig. 6-69 *Replacing leaky parts of galvanized pipe.*

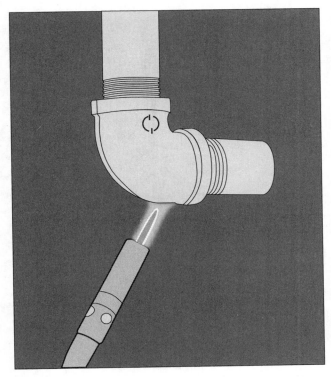

Fig. 6-70 *Heat a stubborn fitting to remove it. Heat only the fitting.*

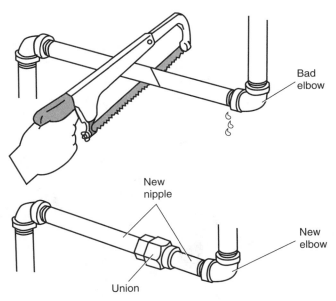

Fig. 6-71 *Fix a leaky galvanized fitting with two nipples and a union.*

This way, you can use a fitting, two short nipples, and a union for the repair. Cut the shortest length of pipe, as shown. Then pull the pipe out just a bit and unscrew either the elbow side or the straight side. Next you unscrew the other piece. Use the procedure shown in Fig. 6-69 to reassemble the pipe and complete the repair.

Drain and Vent Repairs

Drain and vent repairs are rare compared to those for supply pipes. However, they do occur. The process is the same for either cast iron or plastic. Repairing a failed horizontal pipe is not very difficult. Repairing a problem in a vertical drainpipe, particularly if it is a cast iron pipe, is usually a job for the professional. The reason is that any part of a vertical drain supports the weight of the pipe above it. Before you do anything, the weight must be supported. This may take special tools, particularly for a cast iron main drain on a multistory building.

To repair a horizontal leak, first cut away the damaged area. Before you cut, check and see if you need to support the pipe. Once pipe support is ensured, make the cut an inch or so away from the leak, just as for supply pipes. Then cut a length of PVC drainpipe of the correct diameter to replace the damaged section. It is an acceptable practice to replace damaged cast iron pipe with PVC sections in almost every code. The PVC pipe is easier to work with.

Complete the repair, as shown in Fig. 6-72, by using two flexible couplings. Slide a coupling on each end of the old pipe well past the joint. Then you put the PVC repair piece in place and slide the couplings into place. Tighten the clamps and the repair is complete.

To repair a fitting in a horizontal run, you can sometimes use rubber or neoprene fittings and clamps. The main consideration is the weight of the pipe. If the weight is allowed to rest on the fitting, it will distort the flexible fitting and cause leaks. If both sections of the pipe can be supported so that no weight is on the fitting, the flexible fittings can be used.

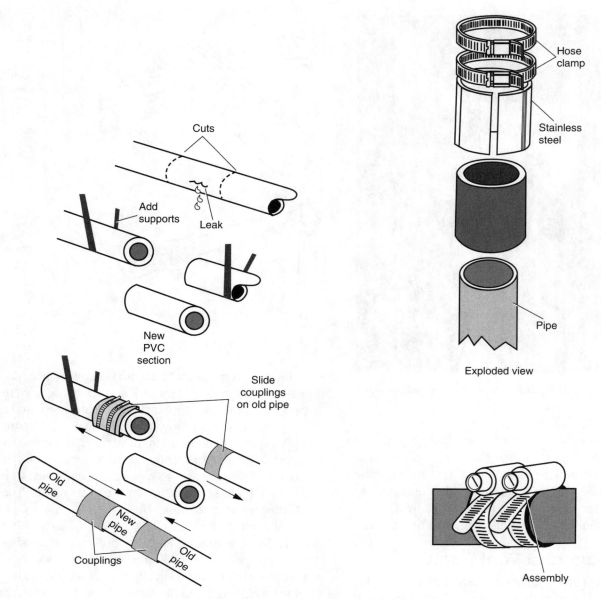

Cuts

Leak

Add supports

New PVC section

Slide couplings on old pipe

Old pipe

New pipe

Couplings

Old pipe

Hose clamp

Stainless steel

Pipe

Exploded view

Assembly

Fig. 6-72 *Replace a bad drain section with PVC and flex couplings.*

7
CHAPTER

Major Plumbing Projects

THE FOCUS OF THIS CHAPTER IS ON MAJOR plumbing projects. Major plumbing projects include several types of jobs. Installing new plumbing for an addition to a house such as a master bathroom is such a project. When you do new plumbing, there are two stages. First, the *rough-in* work involves laying the new supply line and drains. Second, the finish work is done, which includes the installation of cutoff valves and the new fixtures.

Also, replacing old supply lines or drains with new pipes is a major job. This may involve cutting into walls, as in Fig. 7-1, putting in larger pipes, and updating the system.

The installation or replacement of fixtures and appliances is usually a major job as well. Some of these include installing tubs and showers, sinks and lavatories, dishwashers, garbage disposals, and water filters. The installation of toilets might also be a major job, but this was discussed in Chap. 6 and won't be repeated here.

Fig. 7-1 *For major projects, you may need to cut into existing walls to access drains and supply pipes.*

INSTALLING NEW WATER LINES

In this presentation, it is assumed that the floor and walls are framed in, but not covered on both sides. Factors in completing this job are the location of a source for both hot and cold water lines and the joining of the new drains with the main drain line to the sewer. After decisions about where the tie-ins are made and the size and type of pipes are made, the actual pipe work can begin.

The initial pipe work is called *rough-in* work. This is followed by the finish work and installation of the various fixtures and appliances.

Locating a Source

Three sources must be found before you begin running new pipe anywhere. These three sources are for cold water, hot water, and drains. You may not tie into these lines until you have completed everything else, but you must locate the plan for these connections as these locations become your starting points.

Cold water source   First, locate a source for the cold water lines. Probably the first consideration is to examine what size and type of pipe you currently have in the way of supply lines and compare them to what you need for the new plumbing. If you have only a ½-inch line from the meter to your plumbing system, you may need to take it out and install a ¾-inch line there. Refer to the information in Chap. 2 to decide what you need.

Hot water source   Next, you make the same determination as you did for the cold water supply. In some areas, the practice is to locate the hot water heater as near as possible to the meter or cold water source. In other areas, this isn't considered. Further, you should consider the installation of a new hot water heater as well as its location. It might save time, pipe, and money to do this.

For example, if you currently have a 30-gallon water heater and you are planning another bathroom, and perhaps a dishwasher, you may find that a 30-gallon unit is inadequate. If you are going to upgrade to a larger water heater, you can also consider a new location for it.

Drain waste and vent (DWV) factors   Before you start, you must also consider where you will tie in the new DWV units into the main drain for your home. Sometimes you can tie into the main drain inside the home, but often you must do this outside the home. If so, you must be concerned about freeze lines in the soil, pipe sizes and types that meet the codes, and new clean-out plugs. Another factor that may affect your work is the type of pipe in your existing drain. Cast iron pipe may cause you more problems than a tie-in with a PVC line.

Codes and inspections   Finally, you should review the codes that affect the scope of your work and the requirements about inspections for your location. This can affect how long you must wait, and how you plan your work so that you don't lose valuable work time waiting for an inspection. There are also codes on space between fixtures and other details. These will be discussed later.

Running New Pipe

When you start to run new pipe, you should consider where and how to run the pipes, how to start the line, and how to arrange pipes at the desired location so that

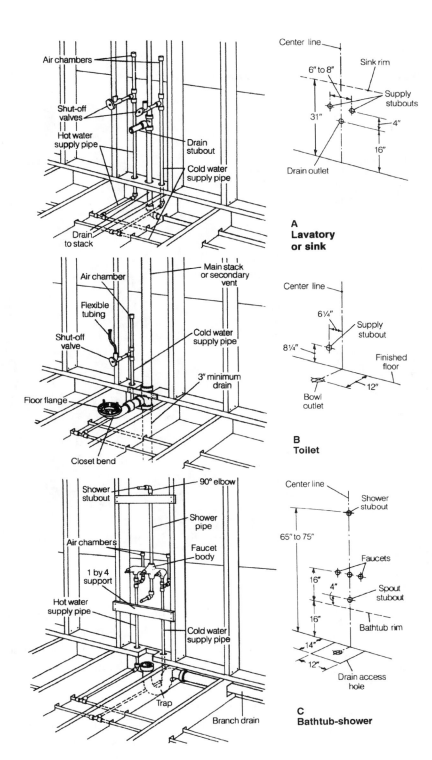

A Lavatory or sink

B Toilet

C Bathtub-shower

Fig. 7-2 *Typical installation dimensions for pipes: (A) Lavatory or sink. (B) Toilet. (C) Tub/shower combination.*

the final fixture can be installed. See Fig. 7-2. New pipes aren't much good if you can't attach the toilet, shower, or lavatory so that it can be used. The last phase is to leave enough pipe for the final connections. The extra length allows the wall to be finished, but leaves short lengths of pipe to which the fixture can be connected. These projecting, but unconnected pipes are called *stubs,* or *stub-outs.* Figure 7-3A, B, and C shows stages of installing a sink from the stubs.

Where to run the pipe The actual location where you intend to tie into the cold water source is of some importance. First, you need easy access to the location if it's possible. You are going to have to turn off the water, cut a pipe, and install some new pipe and fittings to tap into the source. Installing a supply valve might not be a bad idea either.

Next, you need to consider where and how far your new pipe will run. You should avoid running new

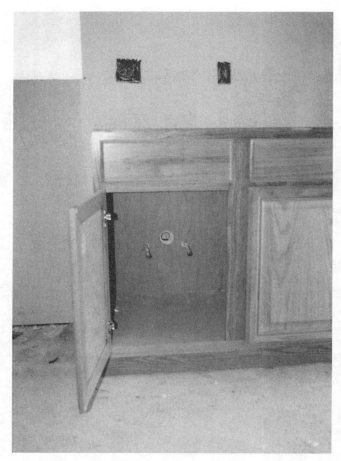

Fig. 7-3A *A cabinet for a kitchen sink is installed over "stubs."*

Fig. 7-3C *The finished sink.*

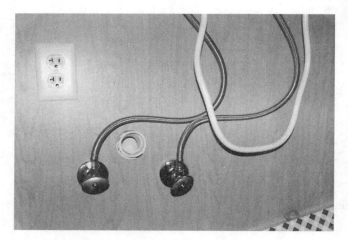

Fig. 7-3B *The sink is mounted first. Then supply valves and lines are installed. The drain is installed last.*

pipe in an area subject to freezing. For example, it is not a good idea to run pipe in outside walls. Anywhere that the pipe is outside the insulation or the insulation is weakened could be the source of a burst pipe. For best results, always run pipe inside interior walls; however it isn't always possible to run pipes through interior walls.

For example, the traditional place for a kitchen sink is beneath a window that overlooks part of the yard. However, it's something you should consider. When you can't do it, then your considerations should include protecting the pipes from freezing.

Starting the new pipe The first step in this job is to locate the point where you will tie into the water source. This will become the starting point for laying the new pipe to the new locations. It isn't necessary to tie into the supply line next to the meter. Connecting at a point nearer to the new installation saves pipe and money.

Also consider the method to use for a tie-in. This can be done with a union, a tee, three nipples, and a valve, as shown in Fig. 7-4, to start the process. If you add a nipple and a cap to the output side of the new valve, you won't get any damage from an inadvertent movement of the valve handle. By installing the tie-in first, you start from a specific location with the source

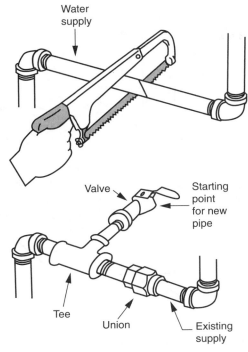

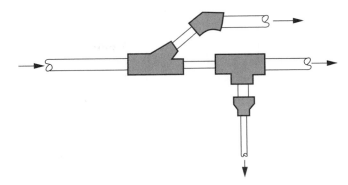

Fig. 7-5 *Fittings let you work pipe to change size and direction.*

Fig. 7-4 *Tie-in for new pipe. Turn off water. Cut pipe, and insert tee and union. Then connect a nipple and valve to the tee.*

and control valve in place. This also lets you turn the water back on in the house for normal use while the new pipe is run. Of course, be sure to cut off the water before you cut into the source.

If you are cutting into galvanized pipe, make the cut at an angle, as in Fig. 7-4. This will allow you to pull one side out a bit and then unscrew one piece more easily.

You begin to lay or run new pipe from the tie-in point. From this point, lay the new pipe in the longest pieces that you can. You connect each new piece to the one you just put in place. You fasten this connection each time by soldering, cementing, or screwing it in place. You may need to cut pipe to an exact length at several points in order to turn corners or branch out to a different location. Place pipe supports as you need them and as you lay each new piece.

Use elbow fittings to change direction, such as when the pipe must turn a corner. Use tees to branch off and supply water to a new location, as in Fig. 7-5. You can also use reducer couplings to go from one pipe size to another. Be sure to remember to include air chambers, dielectric couplings, and electrical grounds in your plumbing where they are needed.

Much of your new work will probably be placed inside of a wall or under a floor. This means that you have to bore holes in the wall studs to place the pipes. When you do this, you should consider four things: (1)

hole size and wall strength factors; (2) pipe lengths that will fit (it's hard to get a long pipe through holes that are inside studs placed 16 inches on centers); (3) support for the pipe to prevent rattles; and (4) protecting the pipe from nails. Remember, after you install the pipe, you will cover over the wall and fasten the covering with screws or nails.

The first factor is normally not much of a problem for supply lines in standard walls at least 3½ inches thick. However, it's a good idea to review it anyway. Every time you cut a hole in a wall stud, you reduce the strength of the wall. As you might guess, there are codes about the size of the hole. Table 7-1 gives information about hole sizes and pipe dimensions. If you are installing ¾-inch supply lines, you can drill the proper hole by using a spade bit or hole saw. Then you must cut the pipe into lengths short enough to push through.

One alternative is to notch the stud as shown in Fig. 7-6. Notching lets you put a longer length into the wall. When the notch is too deep, you must add a "flat" stud, as in Fig. 7-6. The flat stud can be placed on either side of the pipe. Its only purpose is to restore strength to the wall where the original stud has been weakened.

Table 7-1 *Code Factors for Pipes in Walls and Floors*

Wall Codes		
Framing Member	Maximum Notch Size	Maximum Hole Size
2 × 4 Load-bearing stud	7/8″ deep	17/16″ diameter
2 × 4 Non-load-bearing stud	17/16″ deep	2½″ diameter
2 × 6 Load-bearing stud	13/8″ deep	2¼″ diameter
2 × 6 Non-load-bearing stud	23/16″ deep	35/16″ diameter
Floor Codes		
2 × 6 joists	7/8″ deep	1½″ diameter
2 × 8 joists	1¼″ deep	23/8″ diameter
2 × 10 joists	1½″ deep	31/16″ diameter
2 × 12 joists	17/8″ deep	3¾″ diameter

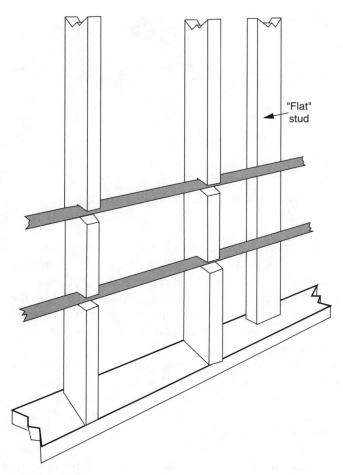

Fig. 7-6 *Install long lengths of pipe in notches. If notches are too deep for codes, add "flat" studs to maintain wall strength.*

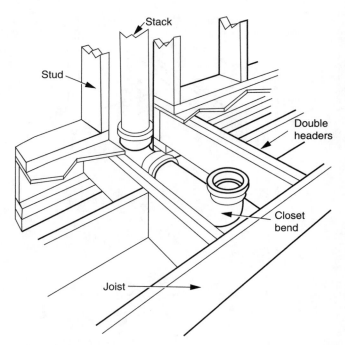

Fig. 7-7 *When you must cut a floor joist, reframe the space, using double headers.*

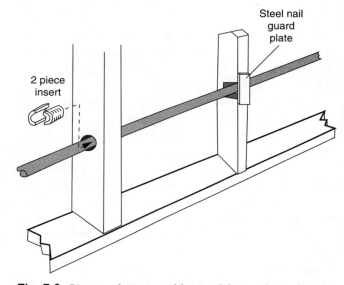

Fig. 7-8 *Pipes can be protected from nail damage by steel inserts and plates.*

You will need a flat stud at every point that you have cut a notch.

If you notch studs in an outside wall, you must allow for the decreased effect of the insulation. The flat stud and the pipe will compress normal spun fiberglass insulation and decrease its effectiveness. Be very sure that the insulation is between the outside and the pipe. Pipe insulation can also help.

Sometimes you may need to cut a floor joist to install plumbing components. When a floor joist is cut, the entire opening should be reframed with double headers, as shown in Fig. 7-7. Be sure that you align both ends of the cut joist before you nail them to the first header.

As pipe or tubing passes through wall openings, the openings can be used as support for the pipe to prevent bangs and rattles. Inserts can be bought or made to fit in the holes or notches. Inserts for round holes can be made from short pieces of a larger pipe, as in Fig. 7-8. You can use steel pipe that can double as nail protection. For support only, you can use almost anything. Short pieces of pipe, wooden shims, or card-

board can all be used. Sometimes support material is governed by local codes, but most requirements only specify some sort of permanent support.

You must also consider damage that can be done by nails and screws. Nails and screws will be used to attach the wall covering after the plumbing is done. Also, pictures may be hung or various decorations added. A nail driven through the stud and into a plastic or thin copper pipe will cause real problems. A small leak inside a wall might not be seen immediately, and it can damage materials from water and possible mildew. It also means you must tear back into the wall to fix the leak. It's very frustrating. When holes are within $1\frac{1}{4}$ inches of the surface of the stud, nail protection is usually required.

To preclude nail damage, put steel inserts into the openings, as in Fig. 7-8, or steel strips over notches, also shown in Fig. 7-8. Nail shields are advisable for plastic and copper pipes and tubing. They are usually not required for steel pipe.

BATHROOM BUILDING CODES

Because bathrooms are complex, building codes may be involved. Codes may designate the types of floors, the materials used, the way things are constructed, and where items are placed. There are usually good reasons for these regulations—even though they might not be obvious. Most cities require rigid inspections based on these codes.

Plumbing

The plumbing is perhaps the most obvious thing affected by codes. Rules apply to the size and type of pipes that can be used, placement of drains, and placement of shutoff valves.

Shutoff valves allow the water to be turned off to repair or replace fixtures. They are used in two places. The first controls an area. It usually shuts off the cold water supply to different parts of the house. A modern three-bedroom house built over a full basement would typically have three area valves—one for the master bath, one for the main bath, and one for the kitchen. Outdoor faucets may be part of each subsystem based on locations, or they may be on a separate circuit. Second, each fixture, such as a hot water heater, toilet, lavatory, and so forth, will have a cutoff valve located beneath it for both hot and cold water lines. Most building codes now require both kinds of valves.

Electrical

Many building codes specify three basic electrical requirements:

1. The main light switch must be located next to the door but outside the bathroom itself.

2. The main light switch must turn on both the light and a ventilation unit.

3. At least one electrical outlet must be located near the basin, and it must be on a separate circuit from the lights. It should also have a *ground fault circuit interrupter* (GFCI).

Ventilation

Ventilation is often required for bathrooms. It is a good idea and has many practical implications. In the past years, doors and windows were the main sources of bathroom ventilation. They consumed no energy but allowed many fluctuations in room temperature. Forced ventilation is not required if the room has an outside window, but most codes require that all interior bathrooms (those without exterior walls or windows) have ventilation units connected with the lights.

Ventilation helps keep bathrooms dry to prevent the deterioration of structural members from moisture, rot, or bacterial action. It also reduces odors and the bacterial actions that take place in residual water and moisture.

Fans are vital in humid clients. They should discharge directly to the outdoors, either through a wall or through a roof, and not into an attic or wall space. Ventilation engineers suggest the capacity of the fan be enough to make 12 complete air changes per hour.

Spacing

Building codes may also affect the spacing of the fixtures such as the toilet, tub, and wash basin. Figure 7-9 shows the typical spaces required between these units. It is acceptable to have more space, but not less.

The purpose of these codes is to provide some minimum distance that allows comfortable use of the facilities and room to clean them. If there were no codes, some people might be tempted to locate facilities so close together that they could not be safely or conveniently used.

Other Requirements

Local codes might require the bathroom door to be at least two doors away from the kitchen. Some locations specify floors to be made of tile or marble, while others mandate tile or marble thresholds. Certain localities insist that a plastic film or vapor barrier be in place beneath all bathroom floors, and that all basin counters have splash-backs or splashboards.

Some specifications might be strict, requiring rigid enclosures on showers or prohibiting the use of glass in shower enclosures. Others might be open, stating the minimum simply to be a rod on which to hang a shower curtain.

FURNISHINGS

Furnishings are the things that make a bathroom either pleasant or drab. They include the fixtures, fittings, and vanity area. The vanity area consists of a lavatory or basin, lights, mirror, and perhaps a counter.

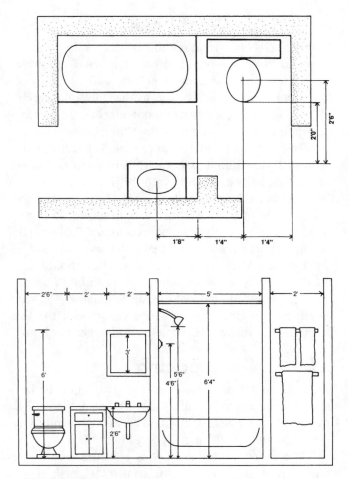

Fig. 7-9 *Typical spacing requirements for bathroom fixtures.*

Fixtures

The term *bathroom fixtures* refers to just about everything in the room that requires water or drain connections, such as lavatories (or basins), toilets, bidets, tubs, and showers. Features to consider for each include color, material, quality, cost, and style.

Generally, the better the quality, the higher the cost. Assuming three grades, the cheapest will not be made to withstand long, heavy use. The difference between medium and high quality will be the thickness of the plating and the quality of the exterior finish.

Toilets, bidets, and some lavatories are made of vitreous china, which is a ceramic material that has been molded, fired, and glazed, much like a dinner plate. This material is hard, waterproof, and easy to clean and resists stains. It is very long-lasting; in fact, some china fixtures are still working well after 100 years or more. White is the traditional color, but most manufacturers now provide up to 16 additional colors. Shopping around can give many insights into colors and features available.

Note in Fig. 7-2 that air chambers are used at all locations for both hot and cold water valves. They are used for the lavatory, the tub, and the cutoff valve for the toilet. Air chambers, as shown in Chap. 3, can be used. However, in many areas, an air chamber is made from the pipe being used. A length of pipe about 12 to 18 inches long is used and is capped at one end. Sometimes plumbers just crimp the end of a copper pipe in two places and fill the crimp with solder. Either way works, but check your local codes first.

Whether the projecting stub is sealed depends on a couple of factors. You may be required to turn on the water and have the whole pipe installation checked for leaks. By doing this, any leaks can be found before the walls are sealed up. This not only is a good idea, but also may be required by the building codes in your area. If this is done, then the stubs must be sealed. There are two ways of doing this. Special stubs, as shown earlier in Chap. 3, can be used. The stubs of pipe can also just be capped. To connect the fixture, the water is turned off at a source valve, and the stub for plastic or copper pipe is cut off with a tubing cutter or saw. With galvanized pipe, the cap can simply be unscrewed.

Drain tie-ins When you add new plumbing fixtures, you must also provide drains and vents. Drain sizes were covered in Chap. 3, and for most projects, they can be 1¼- or 1½-inch PVC pipe. For toilets, you must use larger drains (usually 4-inch diameters). You can use PVC pipe and tie it into cast iron, copper, or other type of pipe in almost every area.

One of the recurring problems in adding new plumbing, however, is adding the proper vent system. Figure 7-10 shows several options. One calculation you must make has to do with the drainpipe diameter. Most codes, as noted in Fig. 7-10, require installation within so many diameters of the trap. This ratio, or *critical distance,* is normally 48 times the diameter of the pipe. That means that the fixture must be within that critical distance to function properly. Thus, the locations for the drain and vent are based on whatever size drainpipe you are using. When you cannot install a fixture within the critical distance, you must use a system called a *back vent.* As in Fig. 7-10, the most important thing to remember about back vents is that they must tie into a vent at least 6 inches higher than the top of the highest fixture in that particular drain system.

REPLACING OLD PIPE

There are two main reasons to replace old pipe with new pipe. The first occurs when the old pipe system is too small and you are plagued by low water flow. The second occurs when the old pipe deteriorates to the point where rust or leaks become problems. There are several things that can cause the second factor.

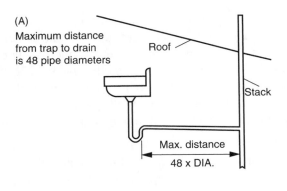

(A)
Maximum distance from trap to drain is 48 pipe diameters

Roof

Stack

Max. distance
48 x DIA.

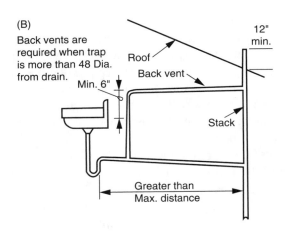

(B)
Back vents are required when trap is more than 48 Dia. from drain.

12" min.

Roof

Back vent

Min. 6"

Stack

Greater than
Max. distance

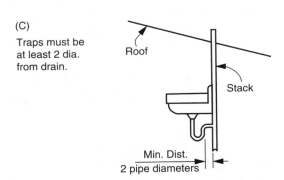

(C)
Traps must be at least 2 dia. from drain.

Roof

Stack

Min. Dist.
2 pipe diameters

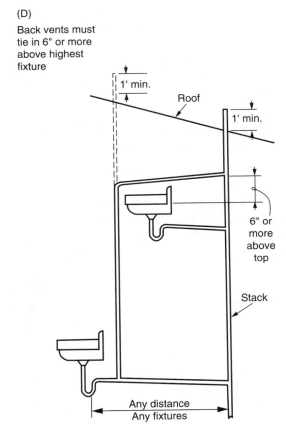

(D)
Back vents must tie in 6" or more above highest fixture

1' min.

Roof

1' min.

6" or more above top

Stack

Any distance
Any fixtures

Fig. 7-10 *Code-required vent and drain distances. (A) Maximum distance from trap to drain is 48 pipe diameters. (B) Back vents are required when trap is more than 48 diameters from drain. (C) Traps must be at least two diameters from drain. (D) Back vents must tie in 6 inches or more above the highest fixture.*

First, if the original pipe was galvanized steel pipe, it may be rusting. If you have rust stains in a sink, this is an indication that the pipe is deteriorating. Further, when a galvanized pipe starts to rust, the water will have a strong taste. The permanent remedy for both of these is to replace the pipe. You may use steel or either copper or plastic pipe, depending on your local codes and your own preferences. Neither plastic nor copper will cause rust problems in the future. Plastic pipe is usually cheaper and easier to work. Copper is stronger and more durable.

Another reason to replace pipe is to place it in a safer location. Suppose that a pipe is located inside an exterior wall. The wall is not insulated, and the pipe has frozen and burst a couple of times. Even if you insulate the pipe, or the wall, the pipe can still freeze in a bad winter. The best preventive method is to move the pipe.

If you live in an area where basements are common, you will want to use interior walls. You can run the new pipe below the floor joists for the ground floor if the basement ceiling is not covered. You can use the joists as bases for attaching pipe supports where you need them. You need only cut small openings near the fixtures and run the pipes vertically through the interior walls.

Some southern areas allow pipes to be run overhead through the attic. This is a convenient method, but it has one major drawback. Attics get very cold in the winter and can freeze the water in a pipe not properly insulated. If you make sure you use pipe insulation as well as cover the pipes with the attic insulation, it is an effective method. Again, it gives a person laying new pipe access to interior walls.

Because this operation is just a form of laying new pipe, you follow all the steps previously outlined. You locate the source, lay the pipe, and stub it out, as before. About the only thing that's different is how you deal with the old pipe. The best rule of thumb is to just leave it in place. This is particularly effective when much of the pipe is between the walls of the house. To remove the pipe would require that long sections of walls be dismantled, the pipe laboriously removed, and the walls recovered. Leaving the old pipe in place would save time, mess, money, and probably a few lost tempers.

INSTALLING NEW FIXTURES AND APPLIANCES

This section focuses on new installation as well as replacing old units with new ones. Once the plumbing is roughed in and stubbed out, and the walls are finished, final installation can occur. For this section, it is assumed that the site is ready. That is, the old unit is removed if it is a replacement job, or stub caps have been removed and supply valves are in place if it is a new installation.

Toilet and supply valve installation was described in Chap. 6. There is no need to repeat those procedures in this chapter. It is also assumed that the water supply to each of the units discussed has been turned off.

The units presented will include the installation of sinks, lavatories, tub/shower combinations, and several appliances. The appliances include dishwashers, garbage disposal units, clothes washers, water heaters, and water filters.

Installing Sinks and Lavatories

There are several locations where sinks and lavatories are used. These include kitchens, bars, laundry and utility rooms, and bathrooms. The hardware and procedure are about the same for all of them. There are a few variations, such as whether the unit is freestanding, wall-mounted, or cabinet-mounted. There are also a wide variety of materials available. Sinks are made of vitreous china or porcelain, enameled steel, stainless steel, porcelain-coated cast iron, and molded stone.

Installing a sink or lavatory involves six factors. Locating the vertical relationship between the trap and the drain is probably the most important of the six factors. The next two concern how the unit is held in place and how it is leveled. The way the unit is connected to the supply lines is another factor. The final two factors are the sealing of the unit against spills and splashes and the attaching of the faucets. All these factors are important for optimum use of a sink or lavatory.

Common factors   Three of the six factors are common to all sink and lavatory installations: locating the vertical distance between the trap and the drain, attaching the faucets, and sealing against spills and splashes. These will be covered first and will apply to all types of units.

Locate the height of the sink. It is usually 31 to 34 inches above the floor. But the single most important factor in the installation process is its relationship to the drainpipe. Figure 7-11 gives a visual description of this relationship. This is an especially important factor in a remodeling or replacement job. If you locate the trap outlet below the drain opening, the unit will drain slowly or simply not drain. The obvious answer to the problem is to raise the height of the sink or basin. This may raise the basin level higher than normal, but it would work better. Sometimes, you can also cut and refit the drain, but that is usually much more involved than just raising the basin.

In either case, there is another adjustment that's possible. The ends (called *tailpipes*) of both sink drains and P traps are usually longer than you need. Thus, you have the option of cutting off part of the tailpipe for an appropriate installation.

To ensure that a sink or lavatory drains properly, include the distance that the bottom of the sink is to be above the floor, how far down the sink drain and trap extend, and the location of the drain opening. As mentioned, this is very important for a remodeling or replacement job. If you are dealing with new installation, you may be able to position the drainpipe opening in a desired location.

Faucet attachment is about the same for all sinks and lavatories. The mechanics are the same whether the sink, basin, or lavatory is located in the kitchen, the bathroom, or the laundry. Two situations are to be considered. The first occurs when you are replacing faucets in a sink, and the second occurs when you are installing both a sink and faucets. When you replace only the faucets, you must unscrew them from underneath the sink, as in Fig. 7-12. A basin wrench is used as shown to reach up behind the sink and unscrew the mounting nuts. Once the nuts are removed, the faucet assembly can be pulled up, mineral deposits removed, and the new unit installed.

If you are installing both new faucets and a new sink, the easiest way is to attach the faucets to the sink

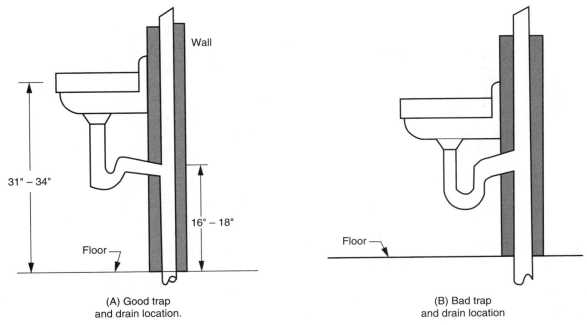

Fig. 7-11 *(A) Good trap and drain location. (B) Bad trap and drain location.*

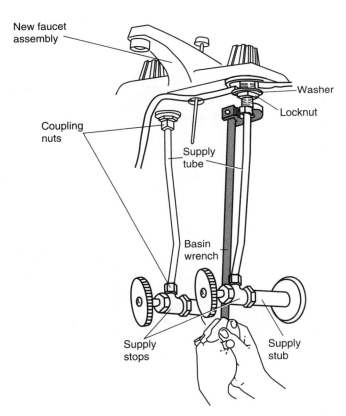

Fig. 7-12 *Old faucets are unscrewed with a basin wrench that can reach up and behind basins.*

before it is installed. Faucets are attached as shown in Fig. 7-13. Some faucets have gaskets under them and others don't. The purpose of the gasket is to prevent water from flowing under the faucet and leaking underneath the sink. In either case, it's a good idea to

seal everything with bathroom caulk, including the gaskets.

Sealing the unit is the final common job in sink and lavatory installation. Once the unit is level, you tighten all fasteners. Then you should seal the unit. The area between the sink and the wall should be sealed with bathroom caulk, as shown in Fig. 7-14. Sealing should be done for all types of sinks in all types of locations. If this area is not sealed, water can collect and damage the wall.

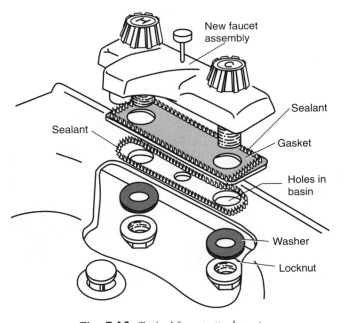

Fig. 7-13 *Typical faucet attachment.*

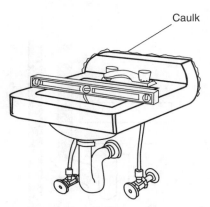

Caulk

Fig. 7-14 *Basins should be leveled and caulked.*

Sink and lavatory mounts The sink mounting is what holds the sink in place so that it can be used. The type of mounting for a sink can affect the appearance of a room and contribute to its cleanliness. There are three basic types of mounts, as seen in Fig. 7-15. These are wall-mounted units, pedestal units, and cabinet units.

The wall-mounted sink allows for movement under the sink and presents an airy and open appearance. This type of mount is ideal for small spaces. It also allows for quick and easy floor cleaning.

At this stage, the rough-out plumbing is in place with the correct spacing and dimensions. Because the sink will be fastened to wall brackets, a support base must be built for the brackets. The distance between the trap outlet and drain opening must a consideration in locating the notched inlets for the brace.

First, decide what is a comfortable distance for a sink to be above the floor. You can do this by measuring other sinks or by checking specification charts. However, remember that the outlet of the trap must slope down to the drain, as in Fig. 7-11. Most sinks and traps have lengthy connecting pipes. The connections are compression fittings, so you can cut the tailpipes to adjust some of these distances. At this point, if the drainpipe is just a stub, you have a lot of wiggle room. If, however, the drain and vent are fixed in place, then you must adjust the height of the sink so that it will drain properly.

Cut inlet notches in the two wall studs nearest to the location of the sink. Then nail a 2 × 4 in the inlets, as

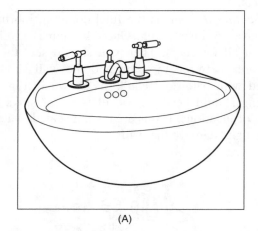

(A)

(C)

(B)

Fig. 7-15 *Three types of sinks or basins: (A) wall-mounted, (B) Pedestal, and (C) cabinet-mounted.*

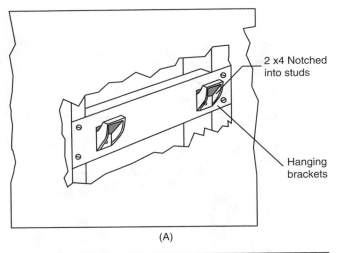

(A)

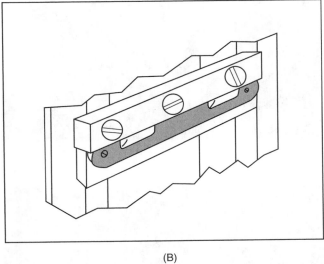

(B)

Fig. 7-16 *Wall-mounted basins are hung on special brackets.*

for best appearance. A material with a pleasing appearance, such as chromed brass or white plastic, is also a consideration. Finally, caulk the top and sides with bathroom caulk.

Pedestal sinks are much like wall-mounted sinks in appearance. However, the sink is supported by the pedestal, rather than wall brackets. There is also a wall bracket for pedestal sinks so that the sink will not move. These brackets are much lighter than the other brackets.

Considerations for drains, supply valves, faucet installation, and seals are the same as for wall-mounted units. Pedestal units may also have floor connectors. These are sometimes visible. For best appearance, the floor bolts are capped with a decorative cover. The sink can be leveled with shims at the floor, and the edges sealed with bathroom caulk, as shown in Fig. 7-17.

Cabinet-mounted sinks, as shown in Fig. 7-18, are very popular. They offer both visual effects and large areas for toiletries. There are two variations of the cabinet-mounted sink, as seen in Fig. 7-19. As shown, one is installed on an open counter while the other is installed on a cabinet that can be used for storage.

There are also two variations in the way sinks are mounted and sealed to the surface of the cabinet. These two types are used in both bathroom and kitchen

shown in Fig. 7-16A. The brackets are attached after the finish wall is put up. As you can see in Fig. 7-16, there are different types of brackets. You should use the type recommended by the manufacturer of the fixture. Often, the manufacturer will include these with the fixture.

After the finished wall is in place, you should have projecting stubs for the hot and cold water supplies and a drain. Attach supply valves to the hot and cold water stubs, as previously discussed. Then the first step is to attach the wall mounting brackets. Do this by first measuring the distance from the floor or drain that you must have and mark this on one side. Next, loosely attach one side. Then, use a level, as in Fig. 7-16B, to determine where to fasten the other side. Mark this and fasten both sides firmly in place.

Next, you hang the sink in place and level it. Then tighten any fastenings. You attach the drain and supply hoses. Because these are open and visible, you should use the shortest lengths and straightest lines possible

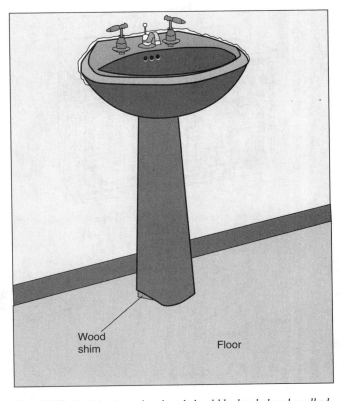

Fig. 7-17 *Both basin and pedestal should be leveled and caulked.*

sinks. Figure 7-20 compares the flush-mounted sink (Fig. 7-20A) with the self-rimming sink (Fig. 7-20B). Both are commonly used. The cabinet top unit is leveled by shims underneath the cabinet. The counter must be hung level on the countertop unit.

Counters often have splash guards, as in Fig. 7-21. Some are one-piece units (Fig. 7-21A) and others have separate pieces with metal or plastic junctions, as shown in Fig. 7-21B. It is best to seal the tops of the splash guards with bathroom caulk. Note that both the flush and self-rimming sinks should also be sealed to the countertop with bathroom caulk.

Molded stone units, as in Fig. 7-22, are another variation of the counter-mounted sink. It features the splash guard, countertop, and basin molded into one solid piece. The units are made of powdered stone mixed with plastic resins and molded to shape. They combine the durability of stone with the molding ability of plastics to provide a luxurious and leakproof area. Molded stone

(A)

Fig. 7-18 *Cabinet-mounted sinks provide work space and storage.*

(B)

Fig. 7-19 *Two types of cabinet-mounted basins. (A) Cabinets provide storage. (B) Counters provide open space beneath the unit.*

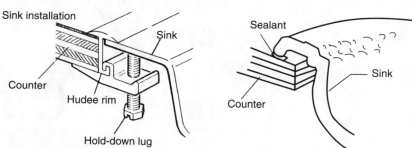

(A) Flush mounted basin (Formica)

(B)Self rimming basin

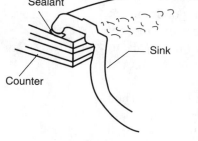

Fig. 7-20 *(A) Flush-mounted basin. (B) Self-rimming basin.*

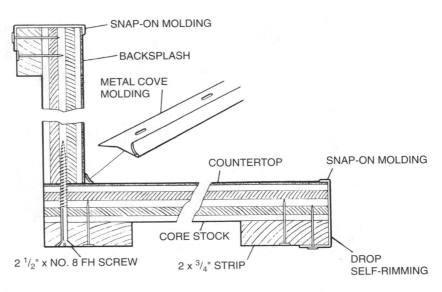

(A) TWO PIECE SPLASH GUARD

Fig. 7-21 *Splash guard details for kitchen and bathroom sinks.*

(B) COUNTERTOP AND SPLASHGUARD MADE AS ONE PIECE.

Fig. 7-22 *A molded stone unit combines the countertop, splash guard, and basin into one solid piece.*

units are more expensive than standard cabinet units, but provide new directions for design. The plumbing on them is just like all other types.

Installing Tub/Shower Units

The installation of a new bathroom adds considerable value to a home. There are three classes of bathrooms: half baths, three-quarter baths, and full baths. The installation of a tub or shower is the determining factor in each case. A half bath has a toilet and a lavatory, but no tub or shower. A three-quarter bath has a shower, but no tub. A full bath has a tub. A full bath may not have a shower, but it will have a tub. Then, of course, there are various luxury concepts, such as those shown in Fig. 7-23A and B.

Before you begin, there are several factors to consider. The first factor is the type of unit, such as tub, tub/shower, sunken whirlpool, or tub/shower with water jets. There are numerous options. The second factor to consider is the material from which the new tub or shower is made. There is the traditional porcelain over

(A)

(B)

Fig. 7-23 *Luxurious concepts for bathrooms. (A) New trends in bathrooms include open space and comfort.* (American Clean Tile) *(B) This bath includes both a spa/tub combination and a glass-enclosed sauna.* (Kohler)

cast iron, enameled steel, molded fiberglass, custom-formed fiberglass, and tile. Tile is considered by some to uncomfortable to sit on, so you should weigh it carefully.

Traditional cast iron is very heavy, and a large unit might weigh well over 100 pounds. In metric weight, that's still a heavy load to lug up stairs, bend around corners, and finally shove into place. Other options include enameled steel, fiberglass, and molded plastics.

Some can be noisy, and others lose heat rapidly. Drawbacks can be offset with added insulation and extra bracing, where needed. All types can provide good service, but cast iron lasts longer and costs more.

Considerations for a combination tub/shower incorporate the processes for both a shower and a tub. Stub dimensions, drain considerations, and seals are essentially the same as those for single-tub or single-shower units. However, your considerations should not be limited to the plumbing details. You may need to consider the overall design, the carpentry work involved, and your wallet.

Preparing a Wall for Tubs and Showers

Walls around tubs and showers must be carefully prepared. This prevents harmful effects from water and water vapor. Walls are finished with tile, panel, and other coverings. Water-resistant gypsum board can be used as a wall base. Special water-resistive adhesives should also be used. Edges and openings around pipes and fixtures should be caulked. A waterproof, nonhardening caulking compound should be used. The caulking should be flush with the gypsum. The wall finish should be applied at least 6 inches above tubs. It should be at least 6 feet above shower bases. See Fig. 7-24. Figure 7-25 shows a tub support board. It is made from 2-inch × 4-inch lumber and is nailed over the gypsum wallboard.

Another technique for preparing walls is shown in Fig. 7-26. In this situation, gypsum wallboard is not used. Instead, a vapor barrier is applied directly over the studs. It can be made of water-resistant sheathing paper or plastic film. A metal spacer is used around the edge of the tub, as shown in the illustration. Next, metal lath is applied over the vapor barrier. The wall is then finished by applying plaster. A water-resistive plaster is used for the tub or shower area.

Special tub and shower enclosures are used. These may be made of metal or fiberglass. See Fig. 7-27A and B. These are framed and braced to manufacturer's recommendations. No base wall or vapor barrier is used.

PANELING WALLS

Walls are often finished with standard-size panels made of wood, fiberboard, hardboard, or gypsum board. Panels made from hardboard, fiberboard, or gypsum are usually prefinished. The surfaces of these panels can be made to resemble wood or tile. Paint, stain, or varnish is not needed. This gives a fast finish with little labor. They can also be covered with wallpaper or plastic laminates. Paint and other materials lend an extremely wide range of appearances. Wood panels are also commonly used. They provide a wide range of wood grains

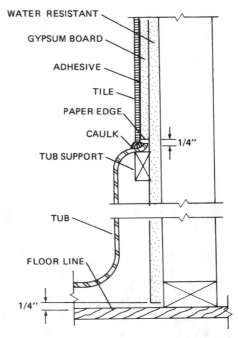

Fig. 7-25 *Tubs are supported on the walls. Note the two layers of gypsum board and the gap used.* (Gypsum Association)

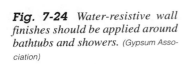

Fig. 7-24 *Water-resistive wall finishes should be applied around bathtubs and showers.* (Gypsum Association)

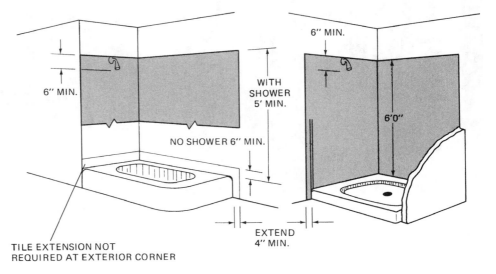

and finishes. Panels may be prefinished. However, unfinished panels are also used. Stains may be applied to customize the appearance of the interior.

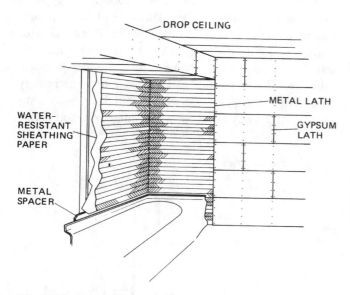

Fig. 7-26 *Vapor barrier and metal lath directly over studs.* (Forest Products Laboratory)

All panels come in standard 4-foot × 8-foot sizes. They are put up with nails, screws, or glue. Special nails and screws are available, colored to match the surface finish. Plain nails are also used. Casing or finish-head nails are recommended. These should be set below the surface and filled. This filling should match the color and texture of the surface.

Panels are available in various materials and range from ⅛ to 1 inch in thickness. As a rule, wood paneling ¼ inch thick needs no base wall. However, the wall is more substantial if a base wall is used, especially with thinner panels. The wall is stronger and more fire-resistant. Many building codes require gypsum base walls. Gypsum drywall is the most common base wall for panels. Some codes require it to have the joints covered. The ⅜-inch thickness is commonly used.

Edges and corners Panel edges and corners are finished in several ways. Figure 7-28 shows several methods. Divider clips are used between panels. To install the panel, a special trick is often used. The edge trim is nailed down first. Then one edge of the panel is inserted into one clip. The second edge is moved away from the wall. Pressure is applied against the edge.

(A)

(B)

Fig. 7-27 *(A) Special tub and shower units are framed and braced by carpenters. (B) This shower stall is made of Fiberglass. No base wall is needed.* (Con Industries)

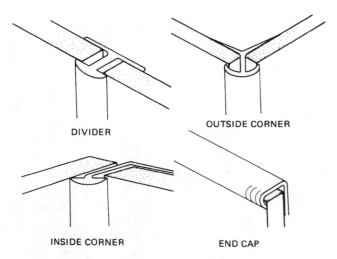

Fig. 7-28 *Joining edges and corners of paneling.*

This causes the panel to bow slightly. The second edge of the panel can then be slipped into the second clip. When the pressure is released, the edge will spring into the clip. The panel will be held flat against the wall. The panel can then be pressed against the adhesives or nailed in place.

BATHING AREAS

A bathing area may be a tub, a shower, or a combination of both. There are many types and varieties of each, and custom units may be built for all types of baths and combination baths.

Bathtubs

Many people like to soak and luxuriate in a tub. Tubs can be purchased in a variety of shapes and sizes (see Fig. 7-29) and can exactly match the color of the other fittings and fixtures in the bathroom. Tubs can also be purchased in shower/tub combinations that match the color of the other fixtures.

Whether to replace an old unit can be an important decision when you are remodeling. Old tubs can be refinished and built in to provide a newer and more modern appearance. Fittings can be changed and showers added, along with updated wall fixtures such as shelves and soap dishes. Sometimes the antique appearance is preferred, in which case reworking is more desirable than replacing.

Sunken tubs may be standard tubs with special framing to lower them below the surface of the floor. They may be custom-made or may incorporate specially manufactured tubs. Before you install a sunken tub, be sure there is room beneath the bathroom floor. When space is not available, the alternative is to raise the level of the floor in the remainder of the room. Of

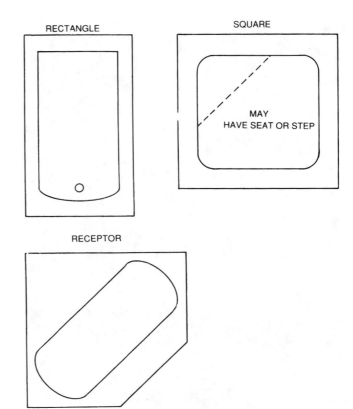

Fig. 7-29 *The three most common tub shapes are rectangular, made to fit into an alcove, square, and receptor. The square and receptor tubs have a longer tub area.*

course, this present considerable complications with existing doors and floors. One compromise is to construct a wide pedestal around the lip of the tub (Fig. 7-30). This pedestal can then become a sitting area or a shelf for various articles, and it can even have built-in storage.

The standard rectangular tub (Fig. 7-29) is 60 inches long, 32 inches wide, and 16 inches high. It is enclosed or sided on one side, but open at both ends and the remaining side. This shape was designed to fit into an alcove, as shown.

Receptor tubs (Fig. 7-29) are squarish, low tubs ranging in height from 12 to 16 inches. Rectangular shapes make them ideal for corner placement. They are approximately 36 inches long and 45 inches wide. Square tubs are similar to receptor tubs in that they can be recessed easily into corners and alcoves. Some have special shelves set into corners, and others incorporate controls in these areas. Square tubs are approximately 4 to 5 feet square, increasing in 3-inch increments. The receptor tub has a diagonal opening while the square tub may have a truly square basin. The disadvantages of a square tub are that it requires a larger volume of water and getting in and out is sometimes difficult, particularly for elderly people.

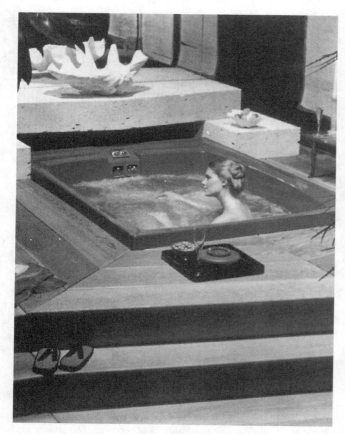

Fig. 7-30 *A pedestal can be built around a tub for many reasons. The pedestal can hide pumps, plumbing, and electrical support units.* (Jacuzzi)

Fig. 7-31 *Tubs can be custom-built to any size and shape. This tile tub features the same color and style of tile for the tub, walls, counter, and floor.* (American Olean Tile)

Fig. 7-32 *The hydromassage action can be combined with more conservative settings to look like a conventional tub.* (Jacuzzi)

Custom tubs (Fig. 7-31) often feature striking use of ceramic tiles. Tiles can be used in combination with metal, stone, and wood. Specially designed custom showers can have stone and glass walls with low tile sides and are used for sitting and storage.

A spa, also called a hydromassage, whirlpool, water jet, or hot tub, requires extra space to house the jet pump mechanism and the special piping and plumbing. Also, wiring is needed to power the motors. These hydromassage units are available in a variety of sizes and can be incorporated into standard tubs. If they are to be installed in addition to the tub, extra space is required. People with smaller bathrooms find that the combination tub/hydromassage unit (Fig. 7-32) is satisfactory.

Specially shaped tubs can be built by making a frame (Fig. 7-33) that is lined or surfaced with a material such as plywood, which generally conforms to the size and shape desired (Fig. 7-33A). Make sure the frame and lining will hold the anticipated weight and movement. Next, tack or staple a layer of fiberglass cloth to the form. Pull the cloth to form the shapes around the corners that are desired. If additional support is desired during the shaping process, corner spaces can be filled in and rounded with materials such as fiber insulation

(Fig. 7-33B). The tub will not need support in the corners, and the fiberglass material itself will be strong enough once completed.

After the cloth has been smoothed to the shape desired (a few seams are all right and will be sanded smooth later on), coat the cloth with a mixture of resin. Tint the resin the color desired for the tub and use the same color for all coats (Fig. 7-33C). Allow the first coat to harden and dry completely. This will stiffen the cloth and give the basic shape for the tub. Next, apply another coat of resin and lay the next layer of cloth onto it. Allow this to harden and repeat the process. At least three layers of fiberglass cloth or fiber will be needed. It is best to add several coats of resin after the last layer of cloth.

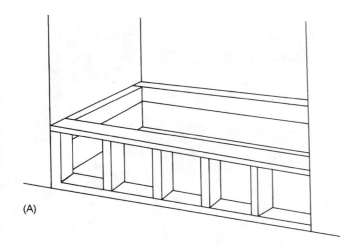

(A)

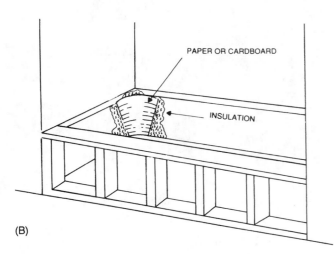

PAPER OR CARDBOARD

INSULATION

(B)

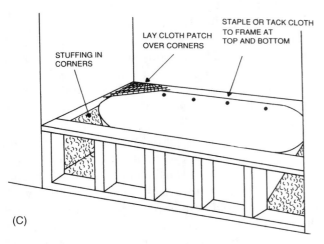

STUFFING IN CORNERS

LAY CLOTH PATCH OVER CORNERS

STAPLE OR TACK CLOTH TO FRAME AT TOP AND BOTTOM

(C)

Fig. 7-33 *Fiberglass tubs can be custom-built to almost any shape. (A) Wooden frame gives dimension and support. (B) Insulation or cardboard defines the approximate shape and contour. (C) A layer of fiberglass is applied. It can be tinted any color to match decor.*

Three layers of glass fiber are generally applied followed by three more coats of resin. The last three coats of resin are sanded carefully to provide a smooth, curving surface in the exact shape desired.

CONSTRUCTION REQUIREMENTS FOR TUBS AND SHOWERS

Several things are needed for installing tubs or showers. These include the supports for the units and the pipes, the pipe work itself, and access for repairs and adjustments. As previously mentioned, the area should be waterproofed as well.

Access Panels

One of the first things to consider should be an access panel, as seen in Fig. 7-34. These may be visible and on hinges as shown in Fig. 7-34A, or hidden underneath counters or in storage areas (Fig. 7-34B). They should be large enough to provide some working room as well as access to the pipes, valves, drains, and diverters. It's a good idea to install cutoff valves in both the hot and cold water supply pipes so that repairs can be made without cutting off water to other areas.

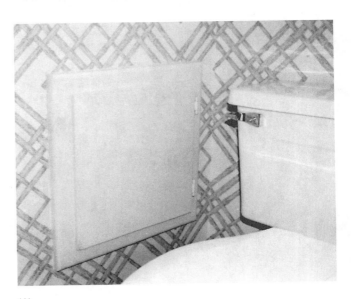

(A)

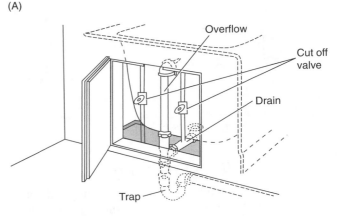

Overflow

Cut off valve

Drain

Trap

(B)

Fig. 7-34 *An access panel should be built into an adjoining wall to provide access to pipes and valves for repairs.*

Pipes and Faucets

Perhaps the first thing to consider is the drain to the shower or tub/shower. This drain must have a P trap, which will extend below the bottom of the unit for several inches. You may need to cut a hole in a floor (even if it's concrete) or build a false floor, as in Fig. 7-35, to achieve a satisfactory drain.

Adding a false floor can be to your benefit. As in Fig. 7-36, the area may be part of a raised luxury feature. The space beneath the tub and surrounding area can also be used. Of course, it can house the pipes, drains, motors, and electrical work. With a little planning, it can also be used for storage.

Pipe and control valve housings for tubs and showers are added before the walls are finished. Both the faucet assembly and the showerhead should be firmly supported, as in Fig. 7-37. Make sure that the dimensions correlate to the manufacturers specifications. The spigot pipe, handles, and actual showerhead are added later.

(A)

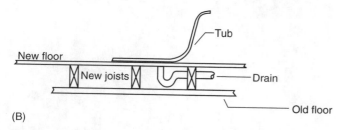

(B)

Fig. 7-35 *To provide space for drains, (A) holes may be cut in floors or (B) a false floor may be built up.*

Tub Framing

The installation of a tub is perhaps one of the most complex steps in plumbing. After the plumbing is roughed in, the final pipes and faucets are installed and braced. Next, the tub framing is done. This includes adding extra floor supports if needed, and putting up the ledgers that

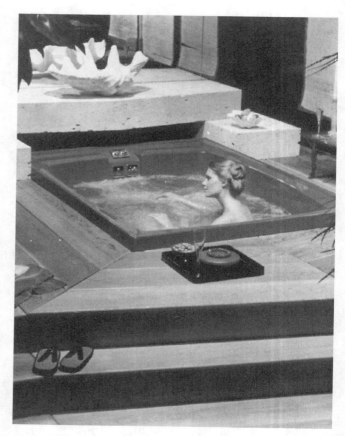

Fig. 7-36 *Sunken tub effect done with built-up floor. This can also house pipes and pumps. It can provide storage space as well.*

will support the tub on three sides. Next, the tub is installed and finally the walls are completed.

Tub framing involves two factors. The first is the consideration of extra weight. Water weighs about 7 pounds per gallon. If a person who weighs 150 pounds wants to soak in 40 gallons of water, the total weight will be over 400 pounds. You may need to add extra floor joists, or nail extra boards to the existing joists to give them more strength.

Tubs are supported partially by the floor, but are also supported and anchored in place with ledgers. Ledgers are 2 × 4 boards nailed to the wall studs. The ledger for the tub is always on the outside of the wall stud. The wall below this point is not covered. The edge of the tub rests on the ledger, and the tub itself becomes the finished exterior surface. The exact location of this support depends upon the size and shape of the tub.

Ledgers around should be used on the back and ends of the tub. These ledgers provide support and hold the tub in place. To frame the ledgers, one should slide the tub in place (Fig. 7-38A), and level it with shims, as shown in Fig. 7-38. The tub has a splash guard type of rim around the edges, as shown in Fig. 7-39A. Mark

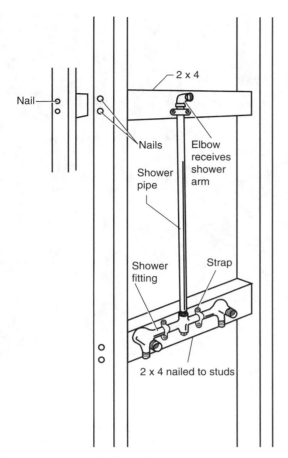

Fig. 7-37 *Nail in supports for tub and shower controls.*

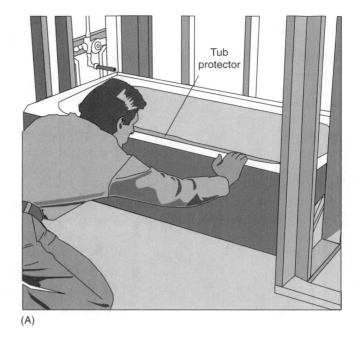

(A)

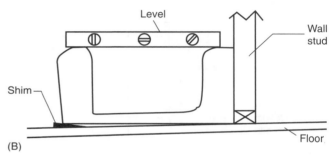

(B)

Fig. 7-38 *(A) Push the tub in place, and (B) level the tub with shims.*

the wall studs and remove the tub. Then measure the distance from the top of the tub rim to the bottom of the tub rim, as in Fig. 7-39B, and lay this distance off below the marks on the studs. This second mark is where the top of the ledger should be. Nail 2 × 4 boards with the tops of the boards at the lower mark, as in Fig. 7-39C.

Install the tub   First lay down a thin coat of dry-set mortar across the area that will be covered by the tub. Lay 1 × 4 boards as shown in Fig. 7-39D to use as slides. They should rest on top of the floor sill on the wall side and on the floor by you. You can rub the tops of the boards with a bar of soap to make the next job a bit easier, but it's not absolutely necessary.

At this point you can also attach the overflow tube. It's a good idea to insulate the tub. This does two things. First, it lets the tub retain heat better. This can prolong the comfort of a hot soak. Second, the insulation will dampen any noise from splashing water.

Wrap insulation around the tub, and then slide the tub up the boards and push it next to the wall. Slide the boards out and push the tub down into the mortar.

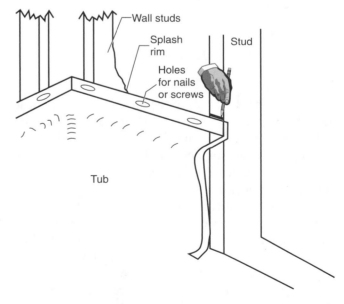

Fig. 7-39A   *(A) Mark the studs.*

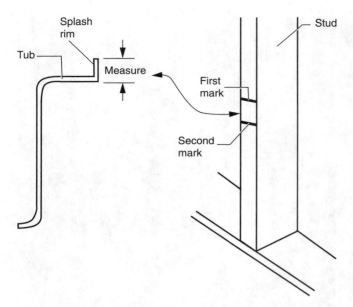

Fig. 7-39B *Measure the splash rim distance and lay this distance off below the first mark on the studs.*

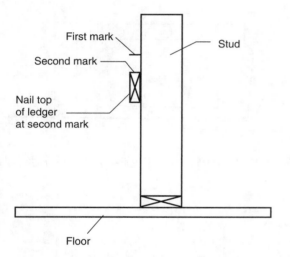

Fig. 7-39C *Nail ledgers at second mark.*

Make sure the edges of the tub firmly seat against the ledgers. Wipe up the excess mortar around the outside edge of the tub. Then nail or screw the tub in place through the mounting holes in the splash rim. It's best to let the tub set for a few hours, but you can begin the next step if time is a factor.

Attach the drain tube and P trap. Connect the trap to the drain, using the compression fittings. Most trap assemblies are longer than necessary, so it's okay to cut off the end of a tube. Just make sure you have enough to seat in the shoulder of the trap.

The next step is to put up waterproof panels around the sides of the tub. As in Fig. 7-25, these panels should overlap the splash rim of the tub and have a small gap

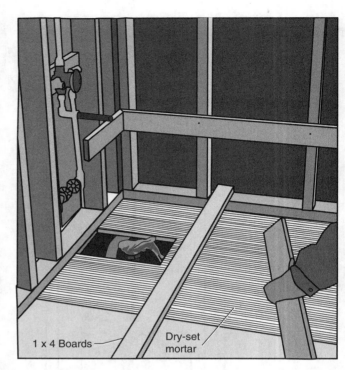

Fig. 7-39D *Lay mortar and 1 × 4 boards for tub slides. Note boards rest on sill of wall.*

at the bottom. The gap is later filled with bathroom caulk. The material used for this is sometimes call *rock board* and is available in most areas. The brand name will often include the word *rock*. It is a material that does not deteriorate when wet or leak water through to the studs. In some areas, a bitumen coat is also added.

The outer wall surface above a tub is often called the *surround*. It should be waterproof regardless of whether a shower unit is installed. The surround surface is added to finish the job. The surround is just the walls around the side and ends of the tub. These must be put up and sealed to shed water from the shower. They may be made of plastic, plastic-coated hardboard, tile, or fiberglass. It's a good idea to include handholds and soap dishes in the sides as well. The surround is discussed in the next section.

Showers

Showers may be combined with the tub (Fig. 7-40) or may be separate (Fig. 7-41). In some smaller bathrooms, showers are the only bathing facility. They can be custom-made to fit an existing space, or standard units made from metal or fiberglass may be purchased. Shower stalls are available with the floor, three walls, and sometimes molded ceiling as a single large unit. When made of fiberglass, they are molded as a single integral unit. When made of metal, they are joined by permanent joints or seams.

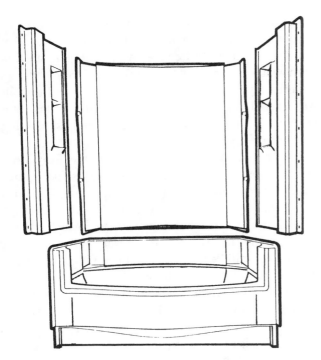

Fig. 7-40 *Combination tubs and showers are perhaps the most common unit. They can be made of almost any material or any combination of materials.* (Owens-Corning Fiberglass)

Fig. 7-42 *A wide variety of accessory fittings such as hand holds, soap dishes, and storage are available for showers.* (NuTone)

Fig. 7-41 *Showers are often separate compartments for increased privacy.*

Standard showers are also available unassembled. This allows a unit to be brought in through halls and doors. The components consist of the drain mechanism, a floor unit, and wall panels.

Manufactured shower units often have handholds, rail ledges for shampoo, built-in soap dishes, and so forth, molded into the walls. Custom-built units can also have conveniences molded into the walls but use separate pieces (Fig. 7-42).

Fiberglass units are usually more expensive than metal ones. Metal units have greater restraints on their design and appearance and are noisier than fiberglass units. Fiberglass should not be cleaned with abrasive cleaners.

Custom-built units are made from a variety of materials including ceramic tile, wood, and laminated plastics. Tile is an ideal material, but it is relatively expensive. The grout between the tiles is subject to stains and is difficult to clean, but special grouts can be used to minimize these disadvantages. Floors for custom-made shower units must be carefully designed to include either a metal drain pan or special waterproofing membranes beneath the flooring.

Using laminated plastics for walls of showers provides several advantages. The material is almost impervious to stains and water, and a variety of special moldings allow the materials to be used. The large size of the panels makes installation quick and easy.

The bottom surface of a shower should have a special nonslip texture; that is, it should be rough enough to prevent skidding, but smooth enough to be comfortable. Neither raised patterns on the bottom of a bathtub nor stick-ons are very good.

Shelves, recessed handholds, and other surfaces in a shower area should be self-draining so that they will not hold accumulated water. Shelves approximately 36 to 42 inches above the floor of the shower are convenient for soap, shampoo, and other items. Handholds, vertical grab bars, and other devices used for support while entering or leaving the shower should be firmly anchored to wall studs.

Separate shower stalls are installed in much the same manner as tubs. Rough-ins and faucets are similar. The bottom of a shower is also called a tub, and it has the same considerations for draining as a tub does. It is also fixed with mortar and nails. The shower bottom will have splash rims and can be finished with metal, plastic, hardboard, or tile surrounds.

INSTALLING APPLIANCES

Many appliances require plumbing as well as electrical connections. These include dishwashers, clothes washers, food/garbage disposal units, and the icemaker or chilled water features on some refrigerators. Other convenience items include water heaters and water filters.

Dishwasher Installation

Dishwashers are great convenience appliances that are now commonly built into new homes. Not too many years ago, that was not the case, and dishwasher installation is a common job attempted by homeowners. Sometimes the homeowner will buy a portable unit that can be later built into the cabinetry. These require no permits or special connections. They use tap water from a faucet adapter and drain into a sink. However, a permanent cabinet installation may be different, so check with your permit office.

If you are considering the purchase of a dishwasher, consider the noise generated in making your decision. The quietest ones have plastic tubs and a heavy insulation around the tub to deaden the sound. Others may have metal tubs with insulation batts wrapped around the tub. The noisiest ones have just metal tubs with no insulation.

Installing a dishwasher in a cabinet isn't very difficult, but there are some key factors to doing it successfully. First, there are some common requirements. Most cities now require an isolated, grounded, or ground fault interrupter (GFI) type of electrical outlet. This means that the outlet is used only for the dishwasher.

If a ground fault interrupter outlet is required, it should be properly grounded to function. This type of outlet is especially effective in providing protection from water-caused electrical problems. Some places require only that a separate circuit be used for the dishwasher and that the outlet be grounded. Ground fault interrupter outlets are readily available at most building supply stores.

Next, you must have an opening for the dishwasher in your cabinet and counter area. The usual requirements are an opening at least 24 inches wide, 24 inches deep, and 34½ inches high. If the dishwasher isn't quite that wide, use molding strips to cover the gaps. The opening must be square, or very nearly so. The dishwasher will be square; if you bend it out of shape, the unit may leak and malfunction.

Dishwashers require only one connection to a water supply. The hot water supply under the kitchen sink is the most commonly used source. The easiest way to tap into this is to buy a tee fitting that will screw on the end of the supply valve you already have. Valves with two outlets are also available. It is a good idea to install a separate cutoff valve in the line to the dishwasher. It may even be a requirement in your area. Figure 7-43 shows how a separate cutoff valve can be used between two connecting hoses to provide a separate cutoff valve for the dishwasher. You can use ¼-inch soft copper lines to connect the dishwasher, but there is one consideration. When the unit is pushed back into the cabinet, the copper line must bend. Care will be needed not to put a kink in the line. The easiest way to connect the dishwasher is with the armored plastic supply lines shown in Fig. 7-43.

Once the electrical and water supplies are done, you must connect the drain. Many new installations drain through the food/garbage disposal unit. This

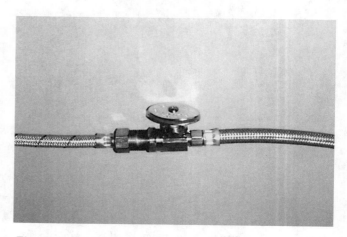

Fig. 7-43 *A cutoff valve for a dishwasher can be installed between two supply lines. Reinforced lines are best for hot water.*

serves two functions. The first is that it is connected to the sink drain. The second is that the disposal unit also serves as an air gap. The air gap prevents wastewater from backing up into the dishwasher. In some areas, a special air gap assembly is required in the installation. The air gap assembly fits into a hole in the sink or at the edge of a drain area.

If you have a disposal unit with a drain attachment, you can just tie the drain from the dishwasher into it and be done. If you are in the process of replacing or adding a disposal unit, get one with a drain attachment.

Once all the connections are made, you are ready to slide the dishwasher into the cabinet and fasten it with screws above the door. If your dishwasher does not have an insulated tub, wrap insulation around it and tape it in place before you slide it into the cabinet. Again, if you use copper tubing, be careful not to kink the line.

Dishwasher repair and maintenance There are several things to watch for and do as routine maintenance and repairs (Fig. 7-44). If the unit will not run at all, check the door. It must close and lock properly to run. There is a switch connected to the lock mechanism that must be fully depressed to work. Check the spray arm for jams and the door opening for obstructions. If this doesn't work, check the plug and outlet to see if you have electrical power. After these, call the repair professional.

Leaks are common dishwasher problems. A typical cause is something distorting the gasket around the door. Food particles, cutlery, and grime buildups are usual causes. Gaskets will also deteriorate with heat and age. Check the gasket also.

If the leak is coming from underneath, the seals may be leaking. The shaft for the spray arm sticks up through the bottom of the tub. There are seals around the shaft to prevent leaks, but if they dry out, the seals shrink. The shrinking seals will not seal properly and will leak. Dishwashers commonly leave a small amount of clean water in the bottom between washes. This serves to keep the seals around the motor shaft from leaking. If you don't use the unit for an extended period of time, the water in the bottom and the seals may all dry out. It usually requires a professional to replace seals.

The most common source of leaks is simply a loose hose. Clamps have a way of working loose. Usually a hose leak can be fixed by just tightening a clamp or nut. If the leak is from a broken or pierced hose, replace the hose.

If the washer doesn't drain, the first thing to do is to check the drain system for clogs. If the dishwasher gets too full, check the float valve in the front corner of the unit. A float valve has a float that rises with the water level. When the water level reaches a certain point, the valve closes and turns off the water supply. These valves sometimes stick. You can check it by simply pulling it up. If it is stuck, it will not come up easily. If this is the case, cleaning is usually the solution.

Finally, it's a good idea to occasionally check the openings in the sprayer arm at the bottom. Bits of food and minerals can sometimes cause buildups that interfere with the operation of the dishwasher. The holes in the arm can be cleaned with a toothpick.

Disposal Installation

Disposal units, as seen in Fig. 7-45, are another appliance gaining in popularity. They are now usually included in new homes but were an "extra" only short years ago. They may be called *garbage* disposal units, or *food* disposal units, or simply *disposal* units.

They are often used as a drain connection for dishwashers, as previously mentioned. If you are purchasing one and you don't have a dishwasher, it's a good idea to buy one with the extra drain connection, just in case. The cost is minimal. This can be capped off if it is not needed. Most new units now come with such drains along with the plugs or caps to block them off if not needed.

To install one, you need an accessible electrical outlet, a separate switch mounted on the wall near the sink, and drain fittings. The disposal is turned on and off by

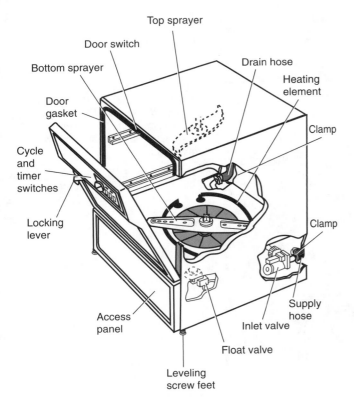

Fig. 7-44 *Dishwasher elements.*

Top sprayer

Door switch

Bottom sprayer

Door gasket

Cycle and timer switches

Locking lever

Access panel

Leveling screw feet

Float valve

Inlet valve

Supply hose

Clamp

Clamp

Heating element

Drain hose

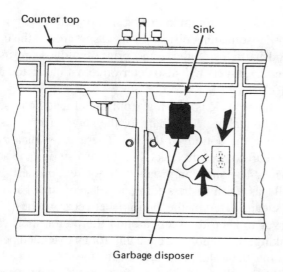

Counter top

Sink

Garbage disposer

Fig. 7-45 *Installation of garbage disposers under the sink. Remember, double-insulated disposers do not require a ground.*

the wall switch. For this reason, the electrical outlet for the disposal must be controlled by the wall switch.

You can install a disposal in a two-basin sink or a single-basin sink. The unit will connect into the regular P trap and drain fittings. Always remember to pull the plug from the electrical outlet when you are working on a unit. If you are installing it, never plug the unit into the outlet until all the plumbing is done.

The first step in disposal installation is to remove the assembly at the bottom of the sink basin where the unit will be. This assembly is usually called a *strainer* or a *basket*. Disconnect the tailpipe and trap, and then remove the basket as seen in Fig. 7-46. Clean away the old plumber's putty and gasket, if any. Then insert the new sink flange. Be sure you have used a sealant and have wiped off the excess. The disposal unit is attached to the sink flange with a special mounting bracket.

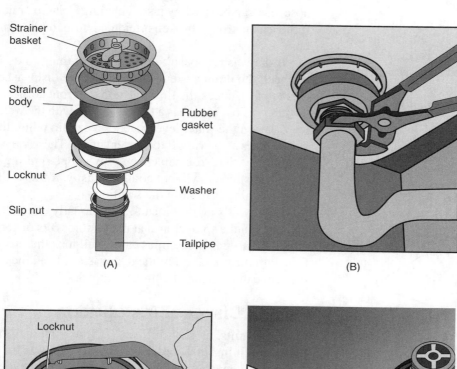

Strainer basket

Strainer body

Rubber gasket

Locknut

Washer

Slip nut

Tailpipe

(A)

(B)

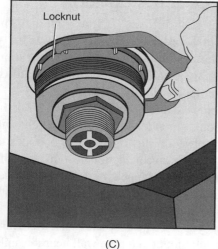

Locknut

(C)

(D)

Fig. 7-46 *To install a disposal unit: (A) Remove the old strainer basket. (B) Remove the slip nut, tail pipe, and trap. (C) Remove the locknut and basket. (D) Scrape the area clean.*

Once the sink flange and mounting bracket have been installed, assemble the drain fittings to the unit. Then insert the unit into the mounting bracket and twist it into position. Tighten it in place with the setscrew.

Once the unit is firmly attached, connect the drain elbow on the unit with the drain. See Fig. 7-47. The drain elbow fits onto the side of the unit at the opening for drainage. It can be swiveled to any angle to allow it to fit a variety of situations. If this is a one-basin sink, then tie the unit into the P trap, as in Fig. 7-48A. If this is a two-basin sink, as in Fig. 7-48B, tie the elbow into the drain. Tighten all the connections on the disposal and on the drain. Run water from the faucet through

Fig. 7-47 *Connect the black disposal drain elbow to the sink drain.*

the disposal to check for leaks. If appropriate, connect the dishwasher drain to the disposal fitting, and tighten it securely. Finally, plug the unit in, turn on the water, and turn on the electricity.

As a last precaution, always turn on the water and let it drain into the unit when you are using it. If you use it "dry," the chopped food particles will not drain properly and the unit will clog and jam.

Clothes Washer Installation

Most new homes typically have built-in connections for a clothes washer, as in Fig. 7-49. These prein-stalled hookups are adequate supply lines; have air chambers to prevent water hammer, P traps, and drains; and are correctly vented. If you are moving into a new home, installation is very simple. You just push the washer partially into place and hook up the two hoses to the hot and cold faucets. Then you place the drain hose into the drain standpipe, and secure it. You need only push the washer into place to compete the job.

If you are going to install a washer without these facilities, you have some plumbing to do. You will have to tap into both hot and cold water supply lines, install hose bib type of faucets, and install air chambers for both hot and cold lines. Then you must tie into an adequate drain and install a P trap and a vent. Figure 7-50 shows the plumbing components for a washer installation. The procedures for the plumbing were presented in Chap. 6.

There are some explanations for these requirements and some exceptions. The first element is the standpipe. The output end of the drain hose must be higher than the water level in the washer. If it is not, it can function as a siphon and drain the washer at the wrong time. Because

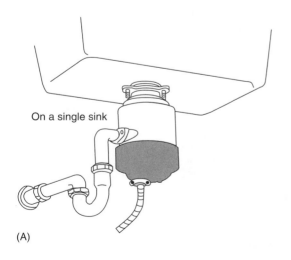

On a single sink

(A)

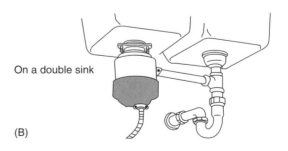

On a double sink

(B)

Fig. 7-48 *Drain configurations for (A) one-basin sink and (B) two-basin sink.*

Fig. 7-49 *A typical new home clothes washer setup. Hot and cold water faucets and the drain are an integral housing. The electrical outlet is also nearby.*

this means both drain and P trap should be installed below the floor. Where this can't be done, one quick solution is to build a false floor and elevate the washer.

The air chambers are also very important. The valves on the washer are solenoid-operated and open and shut almost instantaneously. This is unlike the usual faucet action where the valve is opened much more gradually. For this reason, the surge of water in the pipes is greater than that of regular faucets. You can get a really terrible water hammer when the washer valves are activated. The water hammer will be strong enough to loosen the pipes over time. Thus, air chambers are really very necessary, and they should be of generous proportions.

Once you have installed the necessary plumbing, check it carefully for leaks. Then connect the drain hose and supply hoses as shown in Fig. 7-50. Make sure the water supply hoses have appropriate soft washers and are tightly connected to both the machine and the faucets. If the supply hoses are secured to the machine with clamps, make sure they are very tight. If clamps are used, a double clamp is a good idea. Probably the single most common problem is having the hose connections to the washer loosen and leak. Finally, anchor

the drain hose is smaller than the diameter of the standpipe (usually 1½ inches minimum), air can enter the drain through the standpipe. Thus, the standpipe can function as the vent. However, many places require the washer drain to be vented anyway. A vent is always recommended.

A P trap should be installed below the bottom level of the washer, as in Fig. 7-50. This means below the floor in most cases. If you are installing plumbing for the washer,

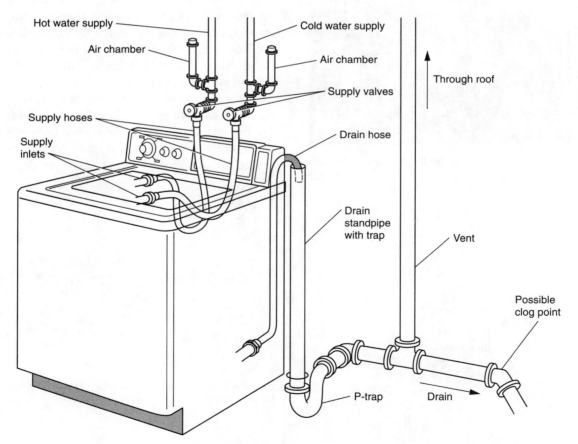

Fig. 7-50 *Clothes washer installation.*

the drain hose to a faucet or a supply line so that it won't slip. Do this with a cable tie, electrical tape, or tie wire.

Once the hoses and drain are installed, turn both supply faucets to the fully open position. Check for leaks around all the hoses and around the stems of the faucets. Then run a test on the washer by putting it through a short wash cycle. Make sure the washer fills rapidly. If it doesn't, then there is an obstruction in the supply or you didn't turn the faucets to the fully open position.

Check to make sure the drain functions adequately. Many washers pump the wastewater out with considerable force and volume. Again, watch for obstructions. Another very common problem is for a residue of clothing fibers and detergent to build up in drains. This problem can be minimized by simply running a couple of wash cycles with hot water, and no clothes, through the drain once each month.

Icemaker and Chilled Water Units

Many refrigerators today are equipped with chilled water dispensers and icemakers. The connection for these is usually a small plastic fitting for ¼-inch tubing on the rear of the refrigerator. This fitting must be connected with a cold water source to enable the unit to work. As shown in Fig. 7-51, the most likely way to do this for copper or plastic pipe is with a self-tapping saddle valve. There are two types of saddle valves. One requires you to drill a hole in the pipe and then install the valve, and the other has a spike that allows you to just screw the handle in to punch a hole in the pipe with the spike. This is the *self-tapping* type. You can't use a self-tapping saddle valve on galvanized pipe. Check for local codes because saddle valves are not allowed in all areas.

You must locate the nearest cold water source first. To connect the refrigerator, a small *feeder* line must be used. If the refrigerator fitting is for ¼-inch tubing, as most are, ¼-inch plastic or copper tubing can be used for the feeder line. Copper is much more durable and is recommended. Some plastics deform and leak when put under pressure. You will also need some sort of valve and appropriate fittings to connect the feeder line to the water supply line.

If your cold water source pipe is plastic or copper, you can use a saddle valve. If you have steel pipe, you must cut the pipe and install a tee and whatever transition fittings you need to attach the feeder line to the refrigerator.

To begin, first shut off the water to the cold water supply pipe you intend to use as the source. Drill the necessary holes in walls for the tubing, and run the tubing

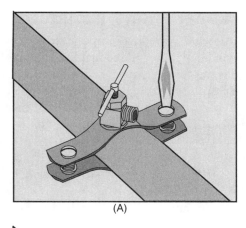

(A)

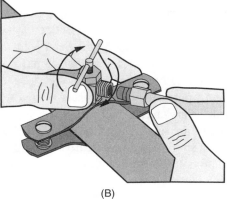

(B)

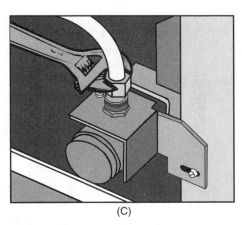

(C)

Fig. 7-51 *To install an icemaker, run flexible copper tubing from a cold water source to the refrigerator. Then (A) attach a saddle valve; (B) Connect the tubing with compression fittings and screw the handle of the valve down; and (C) connect the line to the refrigerator and turn the water back on.*

from the source pipe to the refrigerator. Next, unscrew the handle of the valve and clamp it in place on the source pipe, as in Fig. 7-51A. Once the valve is in place, screw the spike into the pipe to open a passage. Then connect the feeder line to the valve with compression fittings (Fig. 7-51B). Next, connect the tubing to the refrigerator fitting (Fig. 7-51C). Turn the water back on and check the source pipe, valve, and refrigerator fitting for leaks.

Water Filter Installation

Water filters are increasing in popularity and can be easily installed by most homeowners. The most effective type is the cartridge type, shown in Fig. 7-52. The cartridge case can be unscrewed and a new filter element inserted when needed. There are three basic installation patterns for the cartridge-type filter, as shown in Fig. 7-53. These are the single faucet, under-the-counter type (Fig. 7-53A); the single-cartridge, whole-house filter (Fig. 7-53B); and the double-cartridge, whole-house filter (Fig. 7-53C).

The single-faucet filter is installed primarily in a cold water line to improve the taste and quality of the water used for drinking and cooking. A single filter in such a unit may last for several months.

Filters are also installed in the main cold water supply line for a home in order to filter all the water used. These are especially appropriate in older neighborhoods where aging pipes may emit rust and small particles. They are also very effective for rural water systems where sand particles are routinely pumped into the system.

Either a single- or double-cartridge system can be used for a whole-house installation. The advantage of the double-cartridge system is that double-cartridge filters are rated by the smallest size particle that will be removed from the water. Of course, the smaller the particle removed, the more expensive the cartridge is. A cartridge that will stop very tiny particles will typically need replacement more often. Sediment filters effectively reduce amounts of sand, particles, and rust in the water. To reduce bad taste or odor, another type of filter element must be used. (See Fig. 7-54.)

The second type of filter element, also shown in Fig. 7-54, is made from activated charcoal. It also removes sediment, but its primary function is to reduce bad odor or taste from the water. The activated charcoal is effective against chlorine taste and odor in city water. This filter is ideal for an under-the-counter filter and as a second filter in a whole-house system. As you might expect, activated charcoal filters are more expensive than sediment filters.

There are also combination filters that remove both particles and bad tastes. The advantage is that a single filter can do both. With this of type filter, a single filter can be used in a single-cartridge system. The disadvantage is that they are more expensive and require replacement more often.

Either a single- or double-cartridge system can be used for a whole-house installation. The advantage of the double-cartridge system is that this system can provide better filtering of particles and will not need cartridge replacement as often. Further, one cartridge can use a sediment filter, and the second cartridge can hold an activated charcoal filter. This provides the most economical way to reduce both sediment and bad tastes.

There is another advantage of double-cartridge systems. The first cartridge is used as a sediment trap without a filter element. This allows the larger and heavier particles to fall to the bottom of the cartridge. Smaller particles still in the water are filtered out by the second cartridge. This is an economical method of filtering for particles.

To change filter elements, unscrew the cup portion of the filter from the body. You will probably need to use the special wrench for this. Remove the old filter element, and insert the new one into the cartridge cup. Then screw the cup in place by hand. Finally, tighten the cup with the filter wrench, as in Fig. 7-55. Be sure to make it very tight. Read the directions with the unit. Some may state that the O-ring seal on the cup must be coated with some type of grease. If this is the case, oleo or butter does rather well.

Under-the-counter installation To install a cartridge filter underneath a counter (anywhere—bath, kitchen, or laundry) is relatively quick and easy. The unit is most commonly installed between the cold water supply valve and the faucet, as shown in Fig. 7-53. Most filter connections are for ¾-inch pipe threads. Most cold water supply lines under counters are ¼-inch tubing. These two sizes, obviously, don't match. If the filter purchased is

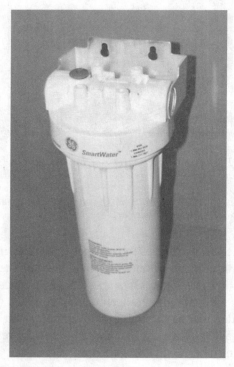

Fig. 7-52 *Cartridge-type water filters are used for both under-the-counter and whole-house installations.*

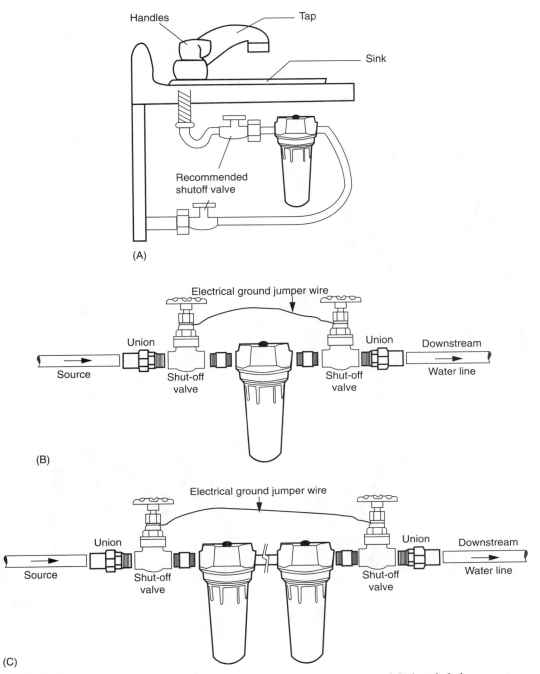

Fig. 7-53 *Cartridge-type filters may be used (A) for a single faucet or (B and C) for whole-house systems.*

specifically for this type of use, the kit will most likely contain all the appropriate fittings. If you purchased only a filter, you will need the transition fittings to change from ¾-inch pipe to ¼-inch tubing.

The easiest way to do this is to use flexible plastic faucet hoses. The standard kitchen or bathroom faucet connection is ½-inch pipe thread. The plastic faucet hose has a ¼-inch tubing connection on one end and a ½-inch pipe connection on the other. Use a ¾-inch nipple, a ¾-inch to ½-inch reducer coupling, and a ½-inch nipple on the intake side of the filter. Then, on the outlet side

(it's marked with arrows showing direction of flow), use a ¾-inch to ¼-inch reducing bushing (be sure the ¼-inch connection is labeled for copper tubing). Wrap all threaded parts with Teflon tape. Remember that standard pipe dope will damage the plastic of the filter.

The under-the-counter filter unit is not connected with rigid pipe. Flexible hoses or tubing is used. Because of this, the filter unit should be supported. A mounting bracket, as seen in Fig. 7-56, can be used, or the homeowner can build a floor unit much like a cup holder.

Whole-house filters The procedure for installing whole-house filters is about the same regardless of the number of filters installed. The water is turned off, and a section of the supply line between the home and the meter, or source, is cut out. How much is cut out depends on the length of the unit being installed. Two filters take more distance than one. See Fig. 7-53. To begin, take apart the filter assembly so that you are working with the top part with the threaded holes.

Filters are usually threaded for 3/4-inch pipe. Transition fittings are required if the supply pipe does not have a 3/4-inch thread. As in Fig. 7-53, once the pipe is cut, nipples and a union will allow you to position the filters as needed. Cutoff valves before and after the filters are also desirable. Remember to wrap all threads with Teflon tape. Do not use regular pipe dope because it will damage the plastic of the filter. Because these filters become part of the permanent piping, they need no separate support.

To complete the job, insert the filter cartridge into the cup or bowl of the unit. Then screw the cup into place with your hands. Finally, tighten the cup very tight with the special wrench, as in Fig. 7-55, and turn the water back on.

Fig. 7-54 There are two types of cartridge filter elements. Left, an activated charcoal filter reduces chlorine tastes and odors as well as particles. The particle filter on the right reduces sand, sediment, and rust particles in the water.

Water Heater Installation or Replacement

Water heaters and hot water are a very important aspect of life today. Water heaters provide the luxury and comfort of hot baths and hot water for household cleaning. They are made large and small, for use in visible

Fig. 7-55 To change filter elements, unscrew cup and remove old filter. Then insert new filter and tighten cup to top.

Fig. 7-56 *Under-the-counter filters can be attached with a mounting bracket.*

locations and for use in out-of-the-way locations. The two most common locations for water heaters are basements and garages. Some electric-powered units are made to fit by kitchen or bathroom counters. These countertop units are ideal for small cabins or apartments.

If you are considering the purchase of a water heater, you must consider the tank size and the heat source. Tank size is usually measured in gallons. A small house with two bedrooms and a single bath will require less capacity than a house with more bedrooms and more bathrooms. A water heater with a capacity of 25 to 40 gallons would be adequate for a house with two bedrooms and one bath. A house with three bedrooms and two baths should have a water heater of a capacity between 40 and 60 gallons. It's always a good idea to buy the larger one when there is little difference in price.

The two most common heat sources in homes are gas and electricity. Each has a tank of water that is heated, but there are some differences typical in the physical makeup of them. Also, gas (either natural gas or propane) is typically the more economical fuel. If you are replacing a unit, check the area your water heater is in to see if you have both energy sources. If not, you should probably use the same heat source again.

If you are replacing a gas water heater, be sure to get a unit for the type of gas you are using. There are two basic types, natural gas and liquid petroleum gas (LPG). The controls for each are different. Most cities have natural gas supplies, while most rural systems use LPG. LPG is often called butane or propane, but LPG has generally replaced them. LPG will vaporize in colder weather than either butane or propane. If you have a "butane" tank in your yard, you probably have an LPG system. Most tanks will have a cover over the main valve. Gas suppliers usually tag the inner surface with their names and phone numbers. If you have any doubts, give them a call. At any rate, do not attempt to install a natural gas heater in an LPG system.

The outward appearance of water heaters is much the same. There is a tank, an overflow valve, a thermostat to control water temperature, and a drain. There is a cold water intake pipe, which should have a cutoff valve

in it, and a hot water outtake pipe. Gas water heaters must also have an exhaust vent.

Gas water heaters, as in Fig. 7-57, feature a gas burner located beneath the tank assembly. Most modern water heater tanks are lined with fiberglass or some plastic material to prevent corrosion. Direct heat on a portion of the tank can damage the lining and reduce the water quality and tank life. The gas water heater tank is usually a circular one with an open space in the middle. The flames of the burner do not touch the tank, but are channeled up through the center opening as shown. The heat from the burner gently heats the water for the entire length of the tank. The unused heat and unburned gas are vented out the top through the vent stack.

Electric water heaters usually are made with a solid tank. Two heating elements are fastened directly into the tank and are in contact with the water, as in Fig. 7-58. Electric water heaters do not require exhaust vents.

Water heater repairs   Water heater repairs are minimal. There are a few things you can do in the way of repairs on water heaters, but it's not always the most economical solution. Water heaters usually last from 10 to 15 years. It's not worth much money or effort to repair a 12-year-old unit. One check of whether the water is hot enough is to simply turn the thermostat temperature control knob back and forth a time or two. Then set it for a slightly higher setting. If lint or oxides

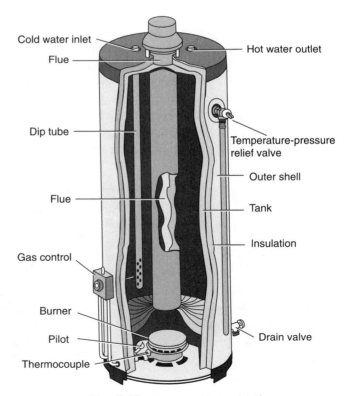

Fig. 7-57 *Gas water heater details.*

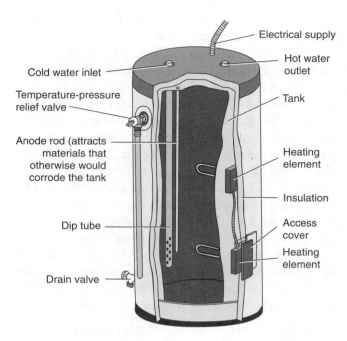

Cold water inlet

Temperature-pressure
relief valve

Anode rod (attracts
materials that
otherwise would
corrode the tank

Dip tube

Drain valve

Electrical supply

Hot water
outlet

Tank

Heating
element

Insulation

Access
cover

Heating
element

Fig. 7-58 *Electric water heater details.*

have impaired the inner works, the movement may clear it up. It the unit begins to heat, the problem is solved.

If a water pipe or a gas pipe leaks, of course, you should fix the leak. If the thermostat or the burner fails on a 12-year-old gas unit, replacement of the whole water heater is probably best.

About the only repair on an old gas unit that is feasible is to replace the thermocouple unit that controls the pilot light. This is a small tube that touches the flame on the pilot. The bimetallic element in it generates a small electric current that holds the gas valve open. If the pilot goes out or the thermocouple fails, the gas is shut off. Thermocouples are readily available and are inexpensive.

There are really only two feasible repairs on an electric water heater. Both concern the heating elements.

To effect either of these, the first step is to cut off the electricity to the water heater.

The first possible repair is to clean the terminals of the heating elements. The reason is that copper forms an oxide when heated. The oxide is literally an insulator and can stop the flow of electricity. Over a period of years, the copper connections can oxidize to the point where they prevent electricity from flowing in the heating element. Once the electricity is turned off, remove the cover over the heating elements. Then just unscrew the connections and clean them. Tighten the connections, turn the electricity back on, and see if it works.

The second possible repair is to replace a faulty heating element. To try this one, turn off the electricity. Unfasten the cover and unscrew the connections to the heating element. Then do a continuity check on the element with an ohmmeter. If the element does not have continuity (it won't conduct electricity), it is bad and you can replace it. You must drain the tank to replace a heating element. Tank draining is covered in this chapter. Once the tank is empty, you simply unbolt the heating element and withdraw it from the tank, as in Fig. 7-59.

If your replacement element does not have new seals, be very careful that you do not damage the old ones. If the old seals are still good, you can reuse them, but seal replacement is the best choice. To finish up, just insert the new heating element and bolt it securely in place. Reconnect all the wires. Refill the tank and turn the electricity back on and it's done. Be sure to check for leaks before closing everything back up.

Removing a water heater The first step in removing a water heater is to disconnect the power source. The gas or electricity must be turned off and disconnected. The gas pipe or wires should be removed from the water heater and twisted to an out-of-the-way location.

The next step is to drain the tank. First, turn off the water at either the cold water intake valve or the main

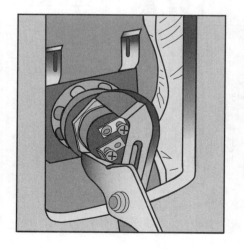

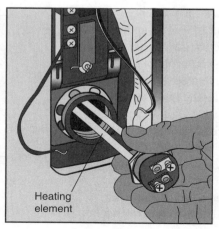

Heating
element

Fig. 7-59 *Electric heating elements can be replaced.*

supply valve to the house. The easiest way to drain a tank is to connect a water hose to the drain and run the water outside or to a basement drain, as in Fig. 7-60. A water heater may also be drained with a bucket. Turn on the drain, fill the bucket, and turn off the drain. Empty the bucket into a sink and repeat the process.

The first few gallons drawn out will drain the water from the pipes and create a vacuum in the tank. The water will not flow very well, but let 2 or 3 gallons flow out. To increase the flow, disconnect the cold water pipe enough to let air into the system. If you don't know which pipe is the cold water pipe, check the top. The word *intake* or *in* should be imprinted there.

Do not disconnect the hot water pipe yet, as it may still have water in it. Once air can enter the tank through the cold water intake, it should drain in just a few minutes. Let the tank drain completely.

The next step is to disconnect the hot and cold water pipes from the water heater. This may be a point where you should bring your installation up to codes. You should have a cold water cutoff valve and some way of disconnecting the pipes from the tank. Disconnects can be effected with unions or with flexible hose fittings, as shown in Fig. 7-61.

If you don't have these, you must cut the pipe to unscrew anything. Remember to cut the pipes at an angle so they can be easily unscrewed. Then install the required valves and connection fittings on the pipes. Most water heater connections are now for ³/₄-inch pipe threads. Water heaters with a capacity greater than 70 gallons may require 1-inch fittings. If your water pipes are not ³/₄-inch galvanized pipe, you will need transition fittings to adapt the water heater to your size and type of pipe.

Also, many codes require gas heaters to be placed in a waterproof pan that is at least 12 inches above the floor. Some places require this only for garage installations. The motivation for this requirement is to raise the open flame of the pilot light above the floor. Many explosive gases are heavier than air and will not reach very high before they are dissipated. By raising the open flame a few inches, some added safety is built into the installation.

Once you have completely disconnected the water heater from the power source, the hot and cold water pipes, and removed the drain hose, you can remove the unit. Lift it up carefully and remove it. If

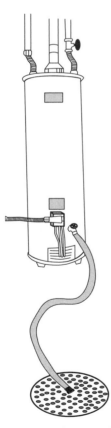

Fig. 7-60 *Drain a water heater with a garden hose.*

Fig. 7-61 *Flexible copper fittings make connections easier and eliminate the need for unions.*

you do not have help, use your knees and not your back to lift the unit. If the unit does not have legs, you can hold the top at an angle and roll it out of the way.

New water heater installation   At this point, the details of the new unit should be decided. You should have obtained the size and type unit you need, and prepared the site for the new heater. This means the pipes are up to code standards; the power connections are nearby; you have the right type of heater for the power source; and stands, vents, and ventilation requirements are met.

Bring the new unit to the location and carefully unpack it. Read the directions with the unit about the installation. Check everything to make sure you have the correct type of unit, pipe sizes, and so forth.

Then connect the pipes and wires as indicated in the directions. Use Teflon tape or pipe dope on all threads. Turn the water back on before you turn the gas or electricity back on. This is so that the heat source will not damage an empty tank. Check everything for leaks.

You can check for gas leaks with a paintbrush and soapy water. A gas leak will make a lot of bubbles. Tighten the connections until there are no visible leaks.

Once the unit has been installed and everything checks out, set the thermostat for a low temperature. The reason is that you don't know how hot this unit will make the water. By using a lower setting, you can let the water heat and test the temperature. This way you can prevent a scalding burn.

8
CHAPTER

Rural Water and Waste Systems

ODAY'S NEW LIFESTYLES HAVE INCREASED the number of people living in rural settings near larger cities. Rural homes are used as weekend and vacation homes. However, such homes are becoming permanent residences as well. The flight from crowded cities and a nostalgic wish for the simple country life have created a new rural family type. These families are not involved with farming or ranching, but often commute or use home office systems. With the movement are problems. The new rural family needs water and waste treatment services previously provided by city utilities. With this new growth, there is a need for better understanding of the water supply and wastewater treatment systems.

Water supply for most rural homes is from water wells on the property. On some older sites, the well may be an old hand-dug one. On newer sites, the new homeowner must have a well dug. The older hand-dug wells are usually at least 30 inches across and up to 40 feet deep. Newer wells dug by boring augers may only be a few inches in diameter, as in Fig. 8-1A. Drilled wells, as in Fig. 8-1B, are dug by drilling rigs and can use several methods. Drilled wells are sometimes called *bores*. Bores may reach a depth of several hundred feet.

Wastewater is disposed in a septic system. Each home has an individual septic tank and dispersal field, as in Fig. 8-2. Effluent from a septic system will be purified as it seeps through soil. One factor often overlooked in early sites was the distance from septic systems.

When the pioneers settled a new land, they may have located the old outhouse too near the water well.

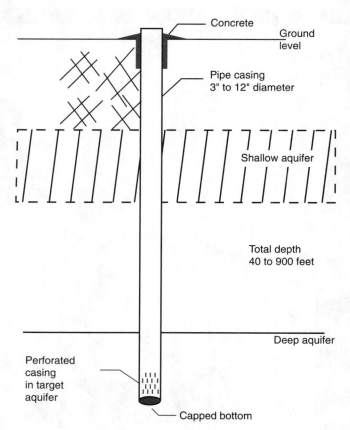

Fig. 8-1B *Typical drilled well configuration.*

The result was eventual contamination of the well from human waste, which caused typhoid fever, hepatitis, and other problems. With the advent of indoor plumbing, the outhouse was abandoned and the modern septic system developed.

The rural homeowner should know the minimum distances needed for the soil to purify the effluent from a septic system. Figure 8-3 diagrams the absolute minimum distances required. These are for sandy and loose soils. If the soil is rich in clay, the distances should be greater.

Water tables or levels determine the depth of a well, as in Fig. 8-1. One should also note that several water tables, or aquifers, may lie at various depths under the surface. Most hand-dug wells tap into the shallowest of these aquifers, while bores may go one or more levels deeper. How deep a well should be depends on the amount of water needed, how dependable the aquifer is at that depth, and the expense. A well drilled to a depth of 500 feet may cost nearly as much as a new house.

Another factor is the quality of the water in an aquifer. Shallow aquifers are more likely to be polluted than deeper ones. For example, when the midwestern states were first settled in the late 19th century, the homes used hand-dug wells. These tapped into the shallowest aquifers. The water was then relatively free from

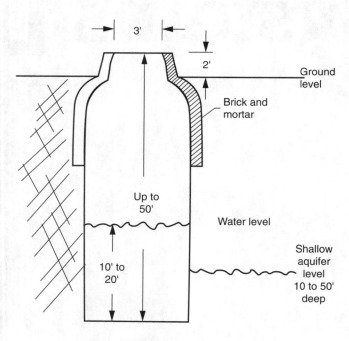

Fig. 8-1A *Typical hand-dug well.*

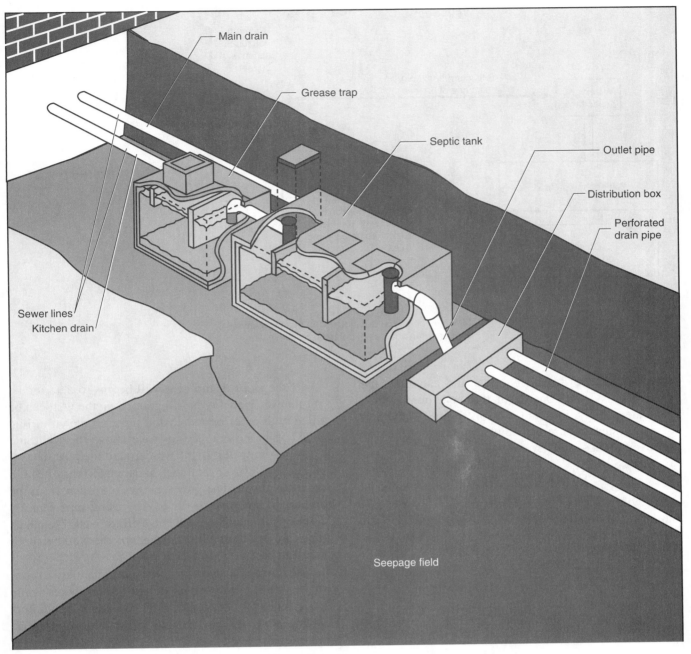

Fig. 8-2 *A modern septic system with drains, three-compartment septic tank and grease trap, and drain field.*

germs or harmful chemicals. Since then, widespread use of nitrate fertilizers, industrial waste dumping, shallow oil fields, and the increase in size and quantity of feedlots have resulted in pollution of these aquifers. In a great many situations, the water is no longer potable (safe to drink) or very good for washing.

This has led to the establishment of *water districts.* These are water systems that are owned by the consumers as shareholders. Typically 15 or more homeowners band together and hire a driller to dig a well deep into an aquifer that is still safe. Then they contract with

others to lay pipelines to the user's homes. They may also treat the water, to add fluorides and purification agents such as chlorine and to *soften* the water. Some water districts provide wastewater systems as well. However, the basic water district provides safe water, and the individual homeowner continues to use the same septic system.

HOME WATER WELLS

Most rural homes probably still rely on their own wells for water. The individual homeowner can add chemicals

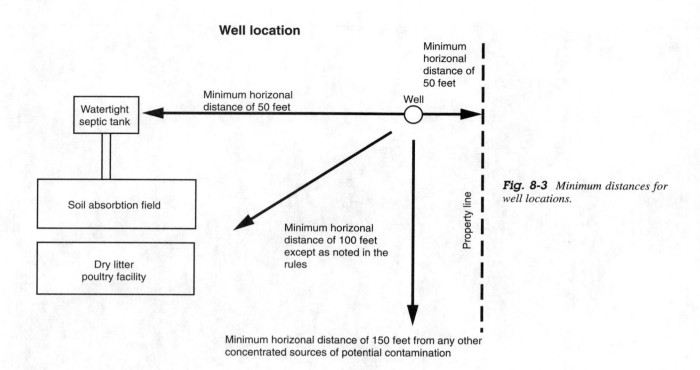

Well location

Minimum horizonal distance of 50 feet

Watertight septic tank

Minimum horizonal distance of 50 feet

Well

Soil absorbtion field

Dry litter poultry facility

Property line

Fig. 8-3 *Minimum distances for well locations.*

Minimum horizonal distance of 100 feet except as noted in the rules

Minimum horizonal distance of 150 feet from any other concentrated sources of potential contamination

(usually chlorine) to kill bacteria and can filter the water. The system consists of the well, pump, tank, and water supply lines and controls. Figure 8-4 shows a basic system. There are numerous types and brand names of pumps, and several well variations.

Well Variations

As previously mentioned, there are bored wells and hand-dug wells. The same basic equipment is used on each, but there are a number of things to know about each. All wells should be sealed to prevent snakes and other animals from entering them. All pipes to and from the well should be protected from freezing and damage. How this is done varies considerably.

Hand-dug wells are dangerous when they deteriorate or are left open. Numerous stories exist about children falling into unprotected wells. Further, the masonry of hand-dug wells, as seen in Fig. 8-5, can deteriorate from weather, exposure, or vehicular traffic. It should be inspected at least annually.

Hand-dug wells must be sealed in two ways. Frequently, the side of the well is pierced to allow the supply pipes to enter the well, as in Fig. 8-5. To seal the well, both top and side seals are needed. The top can usually be sealed with a special lid purchased from a farm supply store. The lid should be weighted, as shown, with at least 40 pounds of weight to prevent the lid from coming off in a high wind. Pierced sides can be sealed with cement, or metal shields.

If the sides or top of a well begins to deteriorate, but the water quality is still good, then the well can be filled with a perforated casing pipe, as shown in Fig. 8-5. The pipe is braced in place above the well, and gravel is dumped into the well around the pipe. Gravel is used as fill for several feet, and then dirt may be used over the gravel. The gravel allows the water from the aquifer to continue to flow. The casing pipe literally converts the hand-dug well into a drilled well. The intake pipes are installed in the casing, and the well becomes safe and useful.

Drilled wells also have distinctive qualities. To construct a well, a professional well driller first bores a hole to a given depth. The driller will know about the aquifers in that area and can provide information on depths, water quality, and estimated costs. At first, the well is just a hole in the dirt. As such, the sides will crumble and cave in. To keep the well open, the well is lined, or *cased,* with pipe. To allow the water to flow into the pipe at the aquifer depth, the bottom part of the pipe must be perforated, as in Fig. 8-1. Perforation can be done in three ways and is usually the choice of the driller.

Special pipe can be purchased that is already perforated. In some cases, the driller may perforate the casing with a cutting torch or perforating tool before the casing is inserted into the well. The third method is to use a special tool with explosive charges to blast holes in the casing. The depth and the amount of the

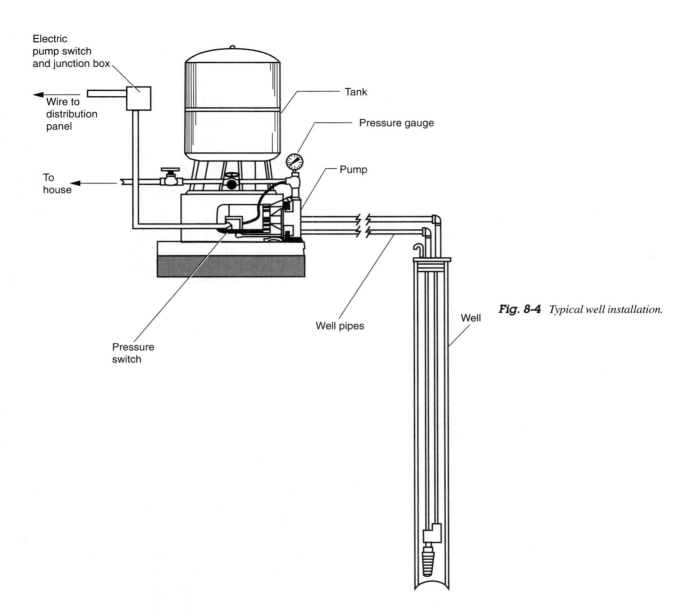

Electric
pump switch
and junction box

Tank

Pressure gauge

Wire to
distribution
panel

Pump

To
house

Well pipes

Well

Fig. 8-4 *Typical well installation.*

Pressure
switch

perforation are determined by the depth and thickness of the aquifer. The casing above the aquifer level is not perforated. The size of the perforation varies as well. If the aquifer is made up of large grains of sand, the perforations can be just small enough to block the grains from entering the well. If the strata particles are small, then the perforations must also be very small to keep the well from silting up. In some locations, this is a problem, and the well must be periodically cleaned.

The diameter of the bore may be as small as 3 inches, but most are 8 to 12 inches in diameter. The larger diameters allow more water to be held in the pipe at one time, and they make pump installation and well servicing easier. Once the well is drilled and cased, the upper part is anchored, or *sealed,* with concrete, as shown in Fig. 8-6. This concrete serves to prevent fuel,

lubricants, or other contaminants from entering the well. There are usually state or county codes that specify the dimensions.

Older wells were lined with steel pipe. Unfortunately, steel rusts and can contaminate the water with particles and foul tastes. Steel casing may last up to 30 or 40 years, but it will eventually cause problems. A new well is an obvious solution, but in some situations, a professional well service can insert a plastic liner into the old steel casing. This blocks the rust from the old casing and restores the water quality. In recent times, well drillers have begun to use heavy plastic PVC casing that doesn't deteriorate as steel does. In shallow wells, the cost is about the same.

Drilled wells are sealed with either the pump or a well cap. This is usually a cast metal or plastic dome

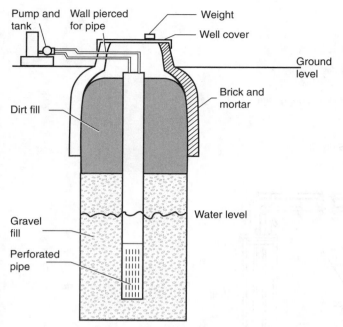

Fig. 8-5 *Dug well with cover and fill.*

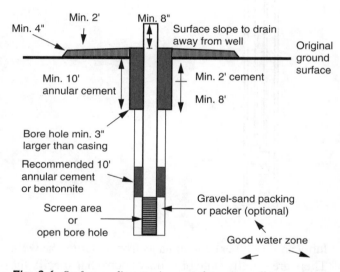

Fig. 8-6 *Surface sealing requirements for water wells.* (Courtesy State of Texas)

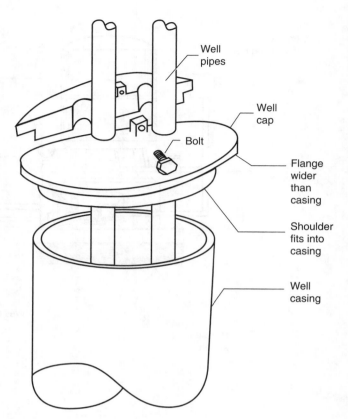

Fig. 8-7 *Well cap is sized for the casing diameter.*

periods of a month or more. However, drilled wells are also susceptible. There are two algae that cause problems. Both black and red algae can grow enough to clog filters and shut off water flow. If you encounter this situation, you must *treat* the well.

To treat the well and get rid of the growths, smells, and odors, first you must remove the cover or cap on the well. Then you pour common household bleach (5% solution of sodium hypochlorite) into the well. About $1/2$ to $3/4$ gallon will treat about 300 cubic feet of water. Then you connect a garden hose to the nearest faucet and insert the end into the well. Position the hose so that the stream of water does not hit the sides of a hand-dug well. In a cased well, it doesn't matter. Then you turn on the water and let the pump recycle the water through the pump and tank for about 30 minutes.

If you have water filters at the well, or in the house, you must physically clean them. You should discard the old filter elements and replace the cartridge cup without a filter element while you recycle the water. You replace the filter elements after you have treated the well.

PUMPS

Two pump systems are commonly used in rural home water supply: jet pumps and submersible pumps. The

that matches the diameter of the well casing, as in Fig. 8-7. The seal will feature a split down the middle so that it can be fitted around the pipe or pipes that bring the water up from the well. This allows the seal to protect the well, but be easily removed for servicing the well.

Well Maintenance

There is little to do to maintain a well except to keep the surface area around the pipe clean. However, wells will develop algae growths that can give the water a rotten-egg smell and a terrible taste as well. This is particularly true for hand-dug wells that are unused for jet pump is by far the most common and the least expensive. The jet system has several variations that make it useful to depths of about 150 feet. The submersible system is designed for greater depths, but can also provide economical service at shallower depths.

Jet pumps use a single impeller to force water through the pipes. Most use a single-phase electric motor to rotate the impeller. Jet pumps have three variations, including the simple siphon, a jet siphon, and a packer jet. Each type requires a valve at the bottom of the intake pipe. This valve is called a *foot valve,* as seen in Fig. 8-8. Without a foot valve, the water in the pump and pipe would drain down to the level of the water in the well. The resulting air pocket between the water level and the pump would break the siphon effect. Without a continuous water stream, the pump will not work.

The foot valve, as in Fig. 8-8, is a rather simple device. A weak spring holds the valve in contact with the valve seat when the pump is not running. When the pump starts, the force of the siphon overcomes the spring and the valve opens, to let the water be sucked up into the intake pipe.

The pump consists of a motor and an impeller unit, and there are several basic locations. Both can be mounted directly on top of the well casing or to one side. Units located to one side may be as much as 45 or 50 feet from the actual well.

Once a jet pump is set up, it must be *primed* to begin pumping. This means the pipes must be filled with water so the impeller can create the basic siphoning action. New units will have specific directions for this. If you are priming an older unit, look for a plug on top of the pump housing. Some units may have a large plug and a smaller one next to it. Unscrew both. The smaller one is a vent and makes pouring the water into the pump and pipes a lot quicker and easier. Pour water into the large opening until it is full. If the system is in good operating condition, once it is full, the water level will stay constant. If the water level in the priming hole continues to drop, there is another problem. The two most likely problems are a leak in the intake pipes or a faulty foot valve. See Table 8-1.

The basic jet pump, or siphon, uses a single suction pipe. This is the same basic principle as when soda is sucked up in a drinking straw. A basic siphon is powered by air pressure. There is about 14 pounds per square inch of pressure at sea level, and this will only allow a siphon action for some 22 feet.

For depths from about 15 to about 90 feet, the siphon jet system is used. This system features two pipes and a special fitting. Water is pumped rapidly through a large pipe across the opening of the smaller pipe inside the special fitting. This creates another siphon below the water level in the well. Once a siphon action is created, it can push water up farther than a simple surface siphon can. For this reason, siphon jet pumps have two pipes from the pump to the fitting. One pipe is almost always smaller than the other. Typical pipe sizes are 1 inch and $1^{1}/_{4}$ inches.

For depths of about 75 to 150 feet, the packer jet system is used. It directs the water flow against the walls of the fitting in several places, to create a series of siphons.

The determining factors in deciding what to use are the water flow rate, the expense of installation and operation, and the depth of the water level. Most of the motors for jet pumps are motors that can be powered by 110 volts or, with a few simple wiring changes, powered by 220 volts for more efficient operation. Simple siphons are less expensive to purchase and install than siphon jets, and so on. Table 8-2 provides a basis for comparisons. As you can see, there is an overlap of a few feet as the depth changes from one type of pump to another. But as you can also note, the deeper a given type of pump is used, the less water flow it provides.

Submersible pumps are used for depths beyond the range of jet pumps. As seen in Fig. 8-9, the submersible pump is a series of impellers within a single case. The whole system is lowered below the water level in the well and pushes the water up. The several impellers act in series to multiply the force. The actual

To pump

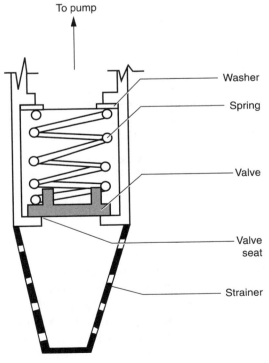

Washer

Spring

Valve

Valve seat

Strainer

Fig. 8-8 *Typical foot valve.*

Table 8-1 Pump Troubleshooting Chart

Trouble	Possible Causes	Remedies
Motor will not run.	1. Disconnect switch is off. 2. Fuse is blown. 3. Starting switch is defective. 4. Wires at motor are loose, disconnected, or wired incorrectly. 5. Pressure switch contacts are dirty.	1. Be sure switch is on. 2. Replace fuse. 3. Replace starting switch. 4. Refer to instructions on wiring. 5. Clean by sliding piece of plain paper between contacts.
Motor runs hot and overload kicks off.	1. Motor is wired incorrectly. 2. Voltage is too low.	1. Refer to instructions on wiring. 2. Check with power company. Install heavier wiring if wire size is too small. See wiring instructions.

Motor runs but no water is delivered.	*1. Pump in a new installation did not pick up prime through:
	  a. Improper priming
	  b. Air leaks
	  c. Leaking foot valve
	*2. Pump has lost its prime through:
	  a. Air leaks
	  b. Water level below suction of pump
	3. Jet or impeller is plugged.
	4. Check valve or foot valve is stuck in closed position.
	5. Pipes are frozen.
	6. Foot valve and/or strainer is buried in sand or mud.
		1. In new installation:
		  a. Reprime according to instructions.
		  b. Check all connections on suction line, air volume control, and jet.
		  c. Replace foot valve.
		2. In installation already in use:
		  a. Check all connections on suction line, air volume control jet, and shaft seal.
		  b. Lower suction line into water and reprime. If receding water level in a shallow well operation exceeds suction lift, a deep-well pump is needed.
		3. Clean jet or impeller according to instructions.
		4. Replace check valve or foot valve.
		5. Thaw pipes. Bury pipes below frost line. Heat pit or pump house.
		6. Raise foot valve and/or strainer above well bottom.

Pump does not deliver water to full capacity. (Also check point 3 immediately above)	1. Water level in well is lower than estimated. 2. Steel piping (if used) is corroded or limed, causing excess friction. 3. Offset piping is too small in size.	1. A deep well jet pump may be needed (over 20 feet to water). 2. Replace with SEARS Plastic Pipe where possible, otherwise with new steel pipe. 3. Use larger offset piping.
Pump pumps water but does not shut off.	1. Pressure switch is out of adjustment or contacts are stuck. 2. Faucets have been left open. 3. Jet or impeller is clogged. 4. Water level in well is too low.	1. Adjust or replace pressure switch. 2. Close faucets. 3. Clean jet or impeller. 4. Check possibility of using a deep-well jet pump.
Pump cycles too frequently.	1. Standard pressure tank is water-logged and has no air cushion. 2. Pipes leak. 3. Faucets or valves are open. 4. Foot valve leaks. 5. Pressure switch is out of alignment. 6. Air charge too low.	1. Drain tank to air volume control tapping. Check air volume control for defects. Check for air leaks at any connection. 2. Check connection. 3. Close faucets or valves. 4. Replace foot valve. 5. Adjust or replace pressure switch. 6. If pressure is below the recommended level, charge the tank. Then check around the valve system with soapy water for leaks.
Air spurts from faucets.	1. Pump is picking up prime. 2. Leak in suction side of pump. 3. Well is gaseous. 4. Intermittent overpumping of well.	1. As soon as pump picks up prime, all air will be ejected. 2. Check suction piping. 3. Change installation as described in manual. 4. Lower foot valve if possible, otherwise restrict discharge side of pump.

*Note: Check prime before looking for other causes. Unscrew priming plug and see if there is water in priming hole.

number of impellers used in a pump varies, but 10 is a fairly common number.

Ten impellers, as you can imagine, provide a powerful force to push the water to the surface. The tradeoffs are that the unit is more expensive than jet pumps, more power is required to operate the unit, and there are more risks to the equipment. As in Fig. 8-9, you can see that the unit is submerged into the well, and its weight is borne by the pipes. Corrosion or bad joints can loosen the unit, and it can sink to the bottom of the well. Special "fishing" tools would be required to retrieve it. Also, the wires to the electric

Table 8-2 Typical Pump Requirements for Siphon Jets and Packer Jet Systems.

Pump	Twin-Pipe Siphon Jet Systems			Packer Systems		
	Min. Well Diameter	Control Valve Setting (psi)	Depth to Jet	Well Diameter	Control Valve Setting (psi)	Depth to Jet
1/2 hp	4″	30	30–40	2″	30	30–40
	4″	30	30–60	2″	30	30–60
	4″	30	60–90	2″	30	60–80
	4 1/2″	30	50–90	3″	30	30–70
	—	—	—	3″	30	50–80
3/4 hp	4″	28	30–60	2″	30	30–60
	4″	29	70–90	2″	32	60–90
	4″	30	90–110	—	—	—
	4 1/2″	28	30–70	3″	27	30–70
	4 1/2″	40	70–110	3″	30	70–110
1 hp	4″	31	30–60	2″	35	30–60
	4″	33	60–80	2″	36	60–100
	4″	35	80–110	—	—	—
	4 1/2″	30	30–80	3″	28	30–80
	4 1/2″	32	80–120	3″	32	80–120

*Simple siphon pumps are used for depths to 20 feet.

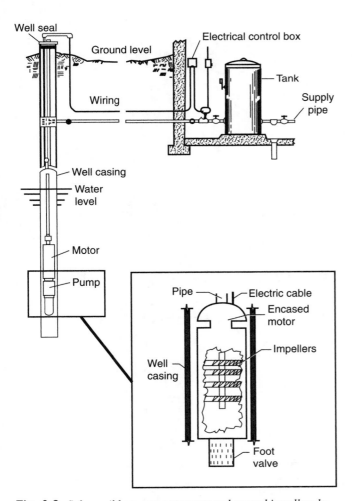

Fig. 8-9 Submersible pumps put motor and several impellers below water level.

motor are submerged, so there is greater potential for electrical problems.

TANKS

Almost all the tanks used in a rural home are one of two types. The first, and oldest, type is an air chamber tank, as seen in Fig. 8-10A and B. The second type uses a thick plastic bladder for the water and is known as a bladder tank.

Bladder tanks These provide several advantages. Both types use an airspace that is compressed by the water as it is pumped into the tank. However, the water in the bladder tank does not touch the metal tank, and thus does not corrode the tank. Further, since the water

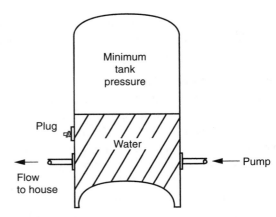

Fig. 8-10A Operation of air chamber tank. At minimum tank pressure, pump turns on.

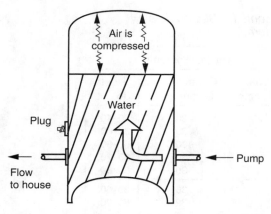

Fig. 8-10B *Pump forces water into tank, which compresses air in top of tank. Pump cuts off at upper limit of switch.*

doesn't touch the metal of the tank, it does not acquire rust particles or tastes from the tank. When the tank is first installed, it is pressurized.

The top of the tank will have an air valve, just as on an automobile tire. Air is pumped into the tank to the recommended pressure (usually 35 pounds per square inch). Then water is pumped into the tank, compresses the air up to about 55 psi, and the system is ready for operation. A control switch is connected into the system so that the pump is switched on when the pressure drops to the minimum, and switched off when the pressure reaches the maximum. Typical pressure range is from about 25 to about 50 psi. This is set by using the pressure gauge on the pump or water line and adjusting the pressure switch to come on or shut off at the lower and upper settings.

The problem with installing a bladder tank is that you also need a source of compressed air. Almost any source, such as a small emergency compressor powered from a cigar lighter in an automobile, can be used. But initial installation requires that the tank have air pumped into it.

Initially, the tank will require little attention. However, after a few months, the tank pressure should be checked with a tire gauge, just as you would check an automobile tire. As the valve stem in the tank ages, it may leak and need periodic replacement.

Air chamber tanks   These tanks are the oldest. The term captive air tank is also used. Most of these types of tanks are galvanized inside and out. There are several variations, however. Some are plastic-lined or coated with enamel. These tanks operate just as the bladder tank does except that the water is not contained in a bladder. The water touches the inside of the tank. As long as the coating inside the tank is undamaged, the steel of the tank will not contaminate the water. However, time and use will usually limit the life of the air chamber tank to less than that of a bladder tank.

Charging an air chamber tank is easy. As you will note in Fig. 8-11A, there is an opening about one-third up the tank. Once the unit is primed and ready to pump, you remove the plug from the opening. Next (Fig. 8-11B), turn the unit on until water comes out of the opening, and then turn the unit off. Put the plug in the hole and make sure it's sealed tight. Remember to use Teflon tape or pipe dope. Then turn the unit on

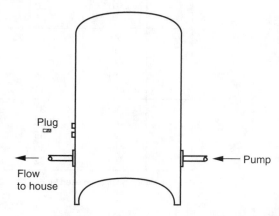

Fig. 8-11A   *Air chamber tank. Initial installation; tank is empty.*

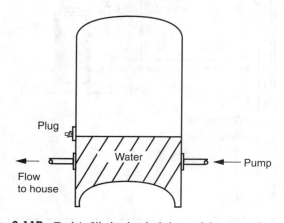

Fig. 8-11B   *Tank is filled to level of plug and the screw plugged in.*

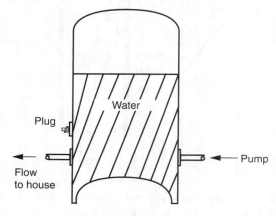

Fig. 8-11C   *Tank should fill two-thirds full for optimal operation.*

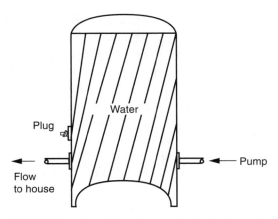

Fig. 8-11D *Waterlogged tank must be drained to plug and recharged.*

again. As the pump fills the tank, the air in the top is compressed to the desired maximum setting. At that point, the pressure switch should cut the unit off. Check for leaks by looking for spurts of water, or damp spots.

Some air chamber tanks have two plugs. The first will be about one-third of the way up the side of the tank. The second may be on top or about two-thirds of the way up the side. You normally don't need these upper plugs to charge the tank. They are helpful when you must drain the tank, however.

You can usually tell how full a tank is by feeling the side. As seen in Fig. 8-12, the water from the well is a different temperature from the air in the top of the tank. In warm weather, the water is colder; but in the worst winter, the water will be warmer. You can almost always tell the water level just by feeling the tank.

It is very common for air chamber tanks to gradually lose the charge and become waterlogged. When this happens, the pump will come on whenever the water is turned on and will run for only a few moments after the water is turned off. You can feel the side of

the tank. If you can't sense any temperature difference, the tank is probably waterlogged. To be sure, loosen the lower plug just a little. If water begins to leak from around the plug, then the tank is waterlogged.

To fix the tank, disconnect the pressure switch by turning off the electricity to the pump. This prevents the switch from operating the pump. Then unscrew the plug on the tank just a little. Next, open a faucet someplace and let the water flow out until the level in the tank reaches the lower plug and stops leaking from around it. Turn off the faucet. Check the level by loosening the plug and checking to see if the water level is at that point.

If the water level is at the correct level, just reseal the opening in the tank and turn on the electricity. The pump will do the rest.

If the tank has two plugs, it's much easier to recharge the tank. Open a faucet somewhere and remove the upper plug. This lets air into the system, and the water in the tank will drain much faster. Loosen the lower plug until it leaks a little. Drain the tank until the water stops leaking from the lower plug. Turn off the faucet, replace both plugs, and make sure they are sealed. Turn the pump back and the pump will do the rest.

Gravity tanks Gravity tanks, as in Fig. 8-13, are giving way to the compact pressure tank units. However, they are still being used in many places. The advantage of a gravity tank is that it will function as long as water is in the tank, even when the electricity isn't working. The disadvantage is with the tower. To achieve a reasonable water pressure, the tank must be located more than 15 feet above the highest floor of the house.

This height also adds to the vertical distance the pump must lift the water. If the water level in the well is 18 feet below the surface and the tank is 20 feet above the ground, then the pump is moving water 38 feet rather than 18.

The tower must also be sturdily constructed. A tower to support a 400-gallon tank must support more than 2800 pounds. Water weighs about 7 pounds per gallon. Then there are problems with insulating exposed pipes, spoiling a scenic view, and so forth.

Supply Lines and Controls

Once the well is drilled and the pump and tank system are operational, water must be carried from the well to the house. This is relatively simple in both thought and deed, but there are some considerations. The first factor is to keep the water in the pipes from freezing in the winter. The pipe is buried below the *freeze line* to protect it. The freeze line (also called the *frost line*) is the typical

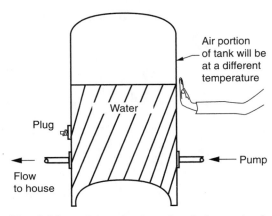

Fig. 8-12 *Feel the side of a tank to find water level.*

Fig. 8-13 *A gravity tank.*

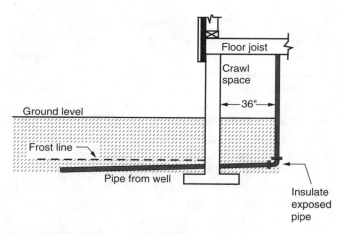

Fig. 8-14 *Pipe from well to home must be buried below the frost line. Pump and tank can be protected in a basement, a well house, or by insulation.*

depth to which the moisture in the soil freezes during cold weather, as in Fig. 8-14. The soil in the south doesn't freeze very deep, as it would in Montana or Maine. The best way to find this depth for your area is to check with your local planning office. There may even be a local code for this.

Pipe type The next factor is the type of pipe to use. Of course, galvanized steel is the strongest, but is more subject to deterioration. Plastic and copper pipes are less prone to lending tastes and so forth. However, there are some other factors. These include animals and equipment. Gophers are common critters in a rural setting. They burrow under the ground and chew through almost anything they encounter. This, unfortunately, often includes soft water pipes and even electrical wiring. Gophers aren't the only ones, but they are good examples.

In an area common to gophers, one should either use galvanized steel pipe or encase a softer pipe in galvanized steel pipe or conduit. Buried electrical wires to the pump unit should also be encased in galvanized conduit.

In a rural setting, there may be heavy equipment traffic. Tractors, plows, large trucks, and so forth may be common around the house. Pipes and wiring should be buried away from the traffic areas if possible. If pipe is buried anywhere near a cultivated area, it should be buried several inches below the deepest depth plowed.

Controls Controls for the system consist of several water valves, at least one faucet, and some electrical controls. As in Fig. 8-15, you should have at least two valves around the tank unit. The pump valve lets you turn off water to the tank to work with the pump. This way, the tank can still supply the house while the work progresses.

The tank valve serves to isolate the pump and tank from the house and yard systems. This can allow repairs to be made on yard faucets or on water supply points in barns and outbuildings. It also serves to isolate pipes for winter closing.

The faucet at the tank allows one to drain the water from the pump, tank, and pipes at that location. This may be needed to effect repairs or to prevent freeze damage.

The house valve, as in Fig. 8-16, can isolate the house for repairs. It is best if the valve has a drain

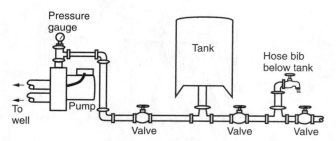

Fig. 8-15 *Desirable plumbing features at pump. Valves allow each section to be cut off for repairs or to drain and recharge tank. Hose bib functions as tank drain or area faucet.*

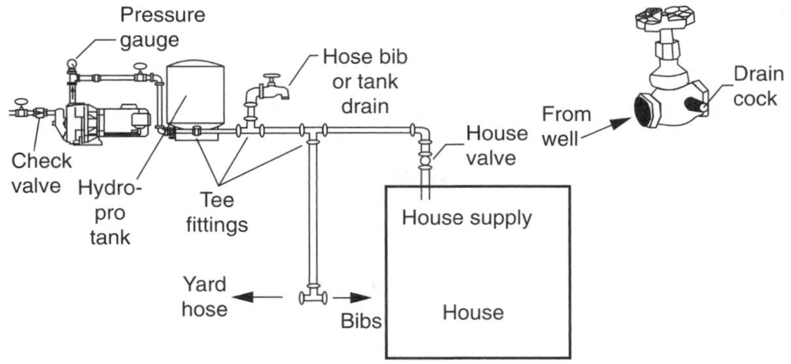

Fig. 8-16 *Typical plumbing layout of supply system from well to house. A bleeder house valve with a drain cock will allow house pipe draining for winter closing.*

cock, as in Fig. 8-16. A valve with a drain feature is called a *bleeder* valve. A bleeder valve allows the water in the house pipes to be drained. This becomes important if the home is to be closed down without heat during freezing weather.

The pump should have several electrical control features. First, as in Fig. 8-17, the pressure switch turns the pump on and off at the desired pressure range. However, as shown, it is a smart idea to have a manual switch and an electrical outlet in the line as well. This would allow you to turn off the electricity at the pump, but still have electricity for a light at that location. A light is very useful if repairs or adjustments are needed at night. A light is also a heat source to ward off freezing problems. The outlet and switch should be housed in a box approved for outdoor use.

The circuit for the pump and well location should be controlled by a separate circuit breaker or fuse. It is unwise to combine the pump unit with another electrical device.

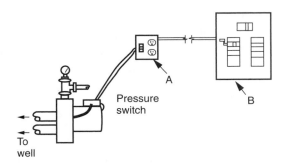

Fig. 8-17 *Desirable electrical features for well pump. (A) Waterproof junction box with on/off switch for pump and 110 V AC always "on" outlet. (B) Separate circuit breaker in main load center.*

WELL HOUSE DESIGN

The pump and tank are often left in the open. However, it is best to build a small *well house* over the unit. A well house protects the components of the system from weather and accidental contact.

The exact size and shape may be selected to match the design of the house. Or, the design may be just big enough to cover the unit, to permit the least visibility. In any case, there are a few factors to consider. The first factor has to do with access to the controls. There are both valves and electrical controls that you must be able to reach. You must also be able to reach the top air pressure valve on a bladder tank, or the plugs used to set the pressure in an air chamber tank.

Perhaps the greatest factor to consider is how the pipes in the well can be pulled out. This may involve pulling out 80 feet or more of pipe. Some well houses are designed so that the roof can be lifted off. Others are built to fold out, while still other well houses can be completely dismantled and set aside by just undoing a few bolts.

When the pipe in the well is pulled out for repairs, it is called *pulling a well*. It isn't done very often because it's a tough job. However, it will need to be done sometime. The pipe is heavy, and the weight is usually beyond the strength of one person. This means using a crane, derrick, or tall tripod of some sort with a lifting apparatus such as a block and tackle. A jury-rigged tripod 8 to 10 feet high with a small block and tackle is the most common rig used by an individual homeowner to pull a well.

When a well is pulled, great care must be taken to prevent the pipe from falling back into the well. If the pump unit is directly on top of the well, then the whole pump unit is pulled out. If the pump is to one side, the pipes between the pump and well are left attached to the well pipe and are used as places to attach the rope or chain used in pulling the well.

The pipe is pulled out as far as possible with the tripod. If steel pipe is used, then it must all be pulled out, or short sections must be unscrewed and removed as it is pulled out.

If the well pipes are polyethylene (PE), they are probably continuous pieces to the siphon jet. The tailpipe down from the siphon jet is probably one piece as well. These pipes can be bent in a wide arc as they are pulled.

Regardless of the type of pipe, the pipe can be pulled out only a few feet with a typical rig. This means unfastening the rope or chain and reattaching it lower down. The way to do this without dropping the pipe back in the well is to use pipe clamps, as in Fig. 8-18. The pipe clamp is firmly attached to the pipe near the top of the well and tied off to something to make sure it doesn't fall back into the well. Then the rope or chain from the block and tackle is unfastened from the top of the tripod. This is then reattached to the pipe clamp, and the next section can be hoisted out of the well. This is repeated until the pipe is out. The process is reversed to reinstall the pipe.

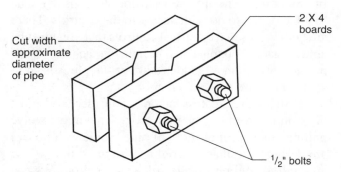

Fig. 8-18 *Make two or three pipe clamps to use when pulling a well.*

Winter Closing

Many rural homes are not occupied year-round. They may be used as weekend or vacation homes and, as such, are closed up when not used. Summer usage doesn't pose many problems. Simply turn off the main valve between the house and tank. Turn off the electricity at the pump, and things are reasonably safe. If the refrigerator isn't going to be used for a while, turn it off and leave the door open to avoid musty odors, and you're about done. Just lock the doors on the way out.

Winter closing is another matter. When the house is not occupied or heated, the temperature in the house will be about the same as the outside temperature. So, the weather inside the house will be freezing cold, just as it would be outside. Anything that contains water may suffer damage. Water expands when it freezes and can exert enough internal pressure to break steel pipe.

The solution is to either drain the water or insulate it from freezing. Points that may suffer damages include toilets, water pipes, water heaters, pumps, and tanks.

Pumps and tanks at a well site can be protected in several ways. Insulation can be wrapped around the exposed tank and the pipes. If the unit is contained in a small building (a well house), protection against freezing may be provided by simply leaving a light on in the well house. If you had an electrical outlet at the well, a light is no problem.

However, for protection from extreme cold, the best solution is to drain everything. This includes water in the tank, the pump, and the pipes. With a faucet at the tank, drainage is no problem. If you included a switch at the pump, the pump can be turned off to prevent it from running.

The house also has water in various places. These include toilets, pipes, and water heaters. First, turn off the water to the house. A bleeder valve, as previously mentioned, will allow you to drain the water from the pipes. Be sure to open at least one faucet to allow air to enter the pipes and help drain the water.

Toilets store water and are very susceptible to freeze damage. The expanding force of freezing water can easily crack the ceramic material of the toilet. Once the main house valve is turned off, flush the toilet and hold the handle down as long as you can hear water run. Then pour a cup of biologically degradable anti-freeze into the toilet bowl. Next pour about a $1/2$ cup of the antifreeze into the toilet tank. The orange type of automobile antifreeze works very well and won't damage the bacterial action of septic systems.

Water heaters should also be drained if winters are extreme. In southern areas where winters are not severe, just draining the water in the pipes and opening the faucet to relieve the pressure are sufficient. In northern climates, water heaters should be drained.

Finally, you should consider the P traps. If you live in a southern region, you probably don't need to do anything. However, for extreme temperatures, you should also pour about $1/4$ cup of antifreeze in each P trap as well.

SEPTIC SYSTEMS

The basic purpose of a septic system is to treat the waste-water from the house so that it poses no health hazards or bad odors. Everything that drains from the house enters the septic system. This includes washwater from appliances, baths, and human wastes.

The water is drained into a large watertight tank where solids are separated from the water. Simple biological action by anaerobic bacterial breaks down the solid wastes, and the fluid effluent is then drained off, where it is absorbed by the soil.

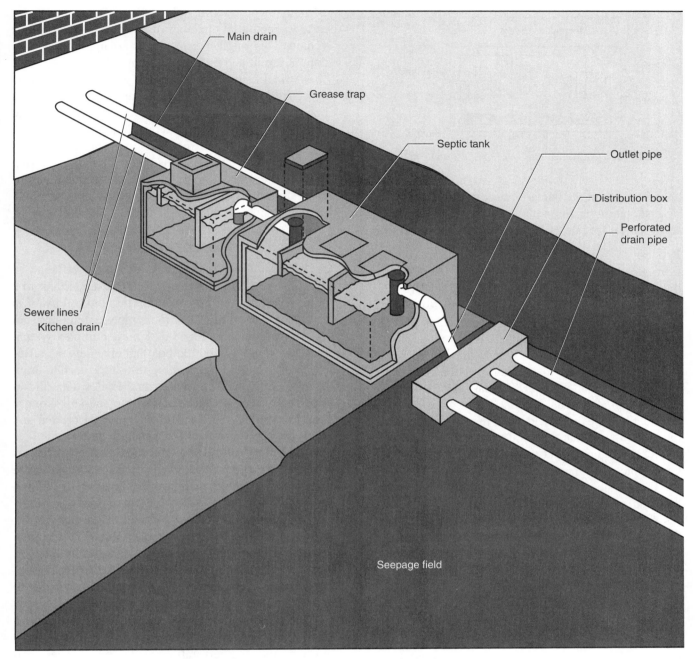

Fig. 8-19 *Separate drain and trap system for kitchen waste water.*

The system consists of a drain, a septic tank, a grease trap, and a drain field, as in Fig. 8-19. Size and location of the system are important factors to consider. Both of these are often specified in local codes. There is a lot of variation in both factors depending upon soil type, population density, proximity to water wells, and so forth.

The septic system must be located so that it will not contaminate a water well (yours, or anyone else's). See Fig. 8-3. It should be located away from the house with minimal visibility. It must also be located where the drain field will be effective.

Septic Tank

The septic tank is a large tank made of something that is waterproof. Most are made of concrete, but heavy plastics are also used. Older systems used one compartment with two surface baffles. Newer tanks, as in Fig. 8-20, have three compartments, all with removable covers. The tank is buried in the ground at a depth where it will not freeze. Septic tanks are usually purchased from and installed by a professional in the business.

When you install a tank, it must be "started." This means that anaerobic bacteria must be added to the inlet

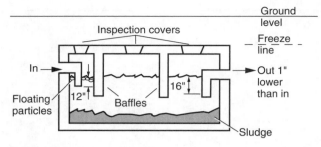

Fig. 8-20 *A typical three-compartment septic tank.*

chamber to start the bacterial decomposition. The septic installation business may provide this or may offer septic *starter* for sale. However, ordinary yeast will work.

Process   In the first compartment, fluids are separated from floating solid wastes. Anaerobic bacteria break down the floating wastes. Anaerobic bacteria derive oxygen from the waste rather than from air. Much of the results of the bacterial action is suspended in the water and passes to the next chamber. Waste materials that are heavier than water fall to the bottom as sludge.

Suspended waste materials flow beneath the baffle and into the outlet chamber. The partially treated water is called *effluent*. The effluent flows from the outlet chamber into the drain field. Note that the outlet is lower than the inlet so that a natural gravity flow is induced. The exact distance that the outlet is lower may vary, but a typical distance is 1 inch.

One sometimes hears the word *cesspool* used to describe a septic system. They are not the same. A cesspool is an obsolete waste system no longer allowed by most codes. It is simply a large hole in the ground. The sides are lined with unmortared stone, and the wastewater seeps directly into the surrounding soil.

Size   Septic tank size is rated in gallons. Typical sizes are 300, 500, 750, and 1000 gallons and larger. In most cases, local codes will specify the size on the basis of the number of bathrooms and bedrooms in a house. For example, a two-bedroom, one-bath home would probably require only a 500-gallon tank, while a three-bedroom, one-bath home would be required to install a 750-gallon tank. Adding another bathroom could boost the requirement to a minimum of 1000 gallons.

Size requirements might also vary with soil type. Loose, sandy soil might lower the capacity required, while dense clay would increase it. Another factor is the type of area. If it is a lakeside home, extra requirements may be needed to protect the purity of the lake water. A home in the middle of a 600-acre farm might require only a small tank.

The tank should be large enough that floating solid wastes remain in the first separation chamber at least

24 hours. The floating solids should not impede new inflow. The outlet chamber of a septic tank should be vented. This might be done with a simple pipe, or a complex vent and baffle system may be used.

A final factor in size depends on frequency of use. A weekend home requires less septic capacity than a home in constant use. You can expect to have the sludge from the septic tank pumped out about once in 5 years. Some homes used infrequently may never need the sludge pumped out. Those used often by a large group may need either a larger tank or more frequent pumping. Pumping is done by a business that specializes in pumping septic tanks.

Drain Field

A *drain field* is a system that moves the effluent, or partially treated wastewater, from the septic tank to a point where it is allowed to drain into the soil. A drain field may consist of one or more pipes extending from the septic tank, as in Fig. 8-3. A more complex system may require a distribution box that channels the effluent into several pipes, as in Fig. 8-3. Pipe used to drain away the effluent is perforated on the sides. This allows the wastewater to drain readily into the soil along a lengthy path. However, solid pipe may be used at any point to carry the effluent to a specific point.

In loose, sandy soil, no special preparation is needed. The pipes are laid in place below the freeze line and are covered. In dense clay, layers of gravel and sand, as in Fig. 8-21, may be required.

The pipes within the drain field will acquire a bacterial layer on the bottom called a *biomat*. This biomat is a spongy, brownish layer of bacterial growth that is normal. It should not block any part of the pipe, but should taper from a thicker layer near the input side to a thinner layer at a distance from the input.

Drain field pipes do get clogged and can back up a sewer line. The most common cause is from tree roots growing into the perforations of the pipes. These can usually be cleared with a power auger. Drain fields also back up when the field is saturated and will not absorb any more water. This is rare with a well-designed system. However, sediments and plant growth may eventually cause saturation. When that occurs, the solution is to install a new drain field.

Grease Traps

Grease traps may be separate from a septic or aerobic system, or operated as part of them. They are similar in construction to a septic tank, but are much smaller. Grease traps treat wastewater from kitchens that carry grease, oil, or grease particles suspended in detergents.

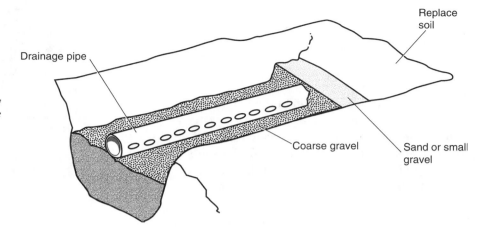

Fig. 8-21 *Sand and gravel may be required for drain fields in dense soil.*

Labels in figure: Drainage pipe · Replace soil · Coarse gravel · Sand or small gravel

These should not be emptied into a septic or aerobic system.

The solution, then, is a separate drain and trap system for kitchen wastewater, as shown in Fig. 8-19. The grease trap removes the grease and oil from the wastewater and then channels it into the septic tank, or into a separate drain field.

The unit should have an inlet and outlet with baffles at each. The outlet should be about 1 inch below the level of the inlet. The baffles keep the floating grease and oil from draining into a drain field or septic system. Bacterial action breaks down most of the grease and oil into sludge, which settles to the bottom. Unbroken grease and oil accumulate between the two baffles. If it stops up, the grease can be skimmed from the surface to unplug the system.

Grease traps should be inspected periodically. Properly designed, they may never need cleaning.

Maintenance of a Septic System

Most well-designed septic systems will function for many years without problems. However, there are a number of things that should and could be done to prolong the service. Of course, there are several things that you should not do, as well.

Things to do   First, use separate gray water drain fields. Adding features to the home such as hot tubs, dishwashers, and clothes washers increases the flow into the system. Wastewater that does not contain human wastes is often called gray water. In some areas, gray water can be drained into a separate field. If you can do this, use a separate drain field for gray water. If you can't do this, either add additional septic tanks in series or install a much larger tank.

Next, spread the use of gray water units out over a period of time. Do one or two loads of wash at once

instead of several. When you use the dishwasher, wait an hour or so before you use something else.

Inspect the system on a regular basis. Check the level of the sludge in the tank. It's best to get the sludge pumped before it backs up the sewer. Walk over the drain field and look for damp spots or odors. Both are indicators of trouble.

Keep records. Have a drawing of the tank and drain field size and location. Keep a record of when you have the system pumped, and review your records once a year.

Don't do these   Don't pour oil, grease, or chemicals into the system. They may kill the bacteria that do the work. Don't plant a garden or trees in the drain field. The roots will clog the pipes, and the vegetables and fruit may not be safe for human consumption.

Don't cover any part of the drain field with anything except grass. Mulch, tree bark, small buildings, or equipment should not be placed over the field. Part of an effective drain field is the evaporation of the water from the soil, so if you cover it with anything, you reduce the efficiency.

Don't use part of the drain field for a road or parking area. The weight of vehicles or heavy equipment can do two things that harm the action of the field. First, the weight can cause the pipes to break or rupture. Second, the weight can compact the soil, sand, or gravel around the pipes. Both factors can greatly reduce the efficiency of the system.

Don't do anything that will add water to the drain field. Don't drain hot tubs, swimming pools, or irrigation systems onto a drain field. These can flood the soil beyond its capacity to absorb the normal flow.

Finally, if you need to add a new drain field because the old one is ineffective, don't tear up the old one. Nature will probably remedy the old field in a few years. Just leave the old one alone for about 5 years, and you can probably use it again. If your soil condition is such that a drain field will only last about 5 years, then by leaving the first one, you can switch back and forth over the years.

AERATED SYSTEMS

Where rural homes spring up around urban areas, the capacity of the soil to absorb and purify wastewater may be diminished. Several homes located on half-acre plots in a small area severely diminish the soil's purification capacity. Dense clay, which has little capacity to absorb water, presents problems to a rural homeowner's wastewater system.

One of the emerging solutions is to use aerobic bacterial action in the tank instead of the passive anaerobic action. These aerobic bacteria must have outside air to thrive. They work faster than the anaerobic bacteria, but require more care. The drain field for an aerobic tank is the same type as the drain field for a septic tank, but may be smaller. Again, local codes and practices govern the size of the tank.

To do this, a more complex system, as in Fig. 8-22, is needed. The tank has the same baffle system, inlet and outlet factors, and may even be the same shape. The major difference is the addition of an electric motor and a mixer/impeller to both break up particles and induce air into the tank. A small tube is used to pump air below the surface of the wastewater while the blades break up the particles and aerate the water.

The advantages of an aerobic system are several. First, the aerobic system provides a more rapid rate of treatment than a septic tank. The aerobic system will work better in dense soil. This factor is particularly important for groups of rural homes on small plots. Finally, it is better suited for areas where water quality is threatened by too many septic systems.

Of course, there are also disadvantages. The first is expense. The motor, aerator, and mixing blades add

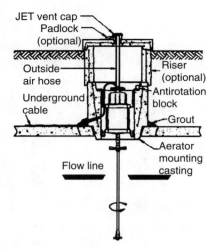

Fig. 8-22 *Aerobic septic systems require the installation of an aerator.*

both moving parts and expense. Power for the electric motor is transmitted through buried wires. The unit requires electricity, and the mechanical parts can break down. Most units do not run continuously, so a timer is also needed to turn the unit on and off, when set. The complexity of the system, compared to a septic system, also demands that it be inspected more often. The operation is more delicate, and heavy loads may not be fully processed.

Again, local codes may either mandate or prohibit aerated systems. Always check. If you consider purchasing an aerobic system, always check for approval by the National Sanitation Foundation (NSF) and the American National Standards Institute (ANSI). Look for a label ANSI/NSF 40 on the unit. This means that the unit meets the consumer standards for these organizations. These organizations check manufacturers, distributors, and installers for correct marketing and installation. Aerator units that meet these standards also have some guarantees that include 2 years of routine inspection and maintenance.

Index

ABOUT THE AUTHORS

Glenn E. Baker is Professor Emeritus of Industrial Technology at Texas A&M University in College Station, Texas. He is the author of a number of books, including *Carpentry & Construction,* Fourth Edition. He lives in College Station, Texas.

Rex Miller is Professor Emeritus of Industrial Technology at State University College at Buffalo and has taught technical curriculum at the college level for more than 40 years. He is the coauthor of the best-selling *Carpentry & Construction,* now in its fourth edition, and the author of more than 75 texts for vocational and industrial arts programs. He lives in Round Rock, Texas.

Mark R. Miller is Chairman and Associate Professor of Industrial Technology at Texas A&M University in Kingsville, Texas. He teaches construction courses for future middle managers in the trade. He is coauthor of several technical books, including the best-selling *Carpentry & Construction,* now in its fourth edition. He lives in Kingsville, Texas.